AF328890

Synthesis Lectures on Biomedical Engineering

Series Editor

John Enderle, Storr, USA

This series consists of concise books on advanced and state-of-the-art topics that span the field of biomedical engineering. Each Lecture covers the fundamental principles in a unified manner, develops underlying concepts needed for sequential material, and progresses to more advanced topics and design. The authors selected to write the Lectures are leading experts on the subject who have extensive background in theory, application, and design. The series is designed to meet the demands of the 21st century technology and the rapid advancements in the all-encompassing field of biomedical engineering

Tristan D. Griffith · James E. Hubbard Jr. ·
Mark J. Balas

A Modal Approach to the Space-Time Dynamics of Cognitive Biomarkers

Springer

Tristan D. Griffith ⓘ
Texas A&M University
College Station, TX, USA

Mark J. Balas
Texas A&M University
College Station, TX, USA

James E. Hubbard Jr. ⓘ
Texas A&M University
College Station, TX, USA

ISSN 1930-0328   ISSN 1930-0336 (electronic)
Synthesis Lectures on Biomedical Engineering
ISBN 978-3-031-23528-3   ISBN 978-3-031-23529-0 (eBook)
https://doi.org/10.1007/978-3-031-23529-0

This Springer imprint is published by the registered company Springer Nature Switzerland AG
The registered company address is: Gewerbestrasse 11, 6330 Cham, Switzerland

*Step into the mystery of the mind.
There are billions of minds, there are billions of
selves. There are billions of worlds and
dimensions.*

—Frederick Lenz

To Tristan and Fiona as they start their new journey in Life together.

About 6 years ago, my wife developed an illness that resulted in a loss of short-term memory. This left her unable to hold a conversation longer than about 10 s. After numerous medications, cognitive tests, and related therapies, her progress was excruciatingly slow. I wondered if my career as an engineering researcher could enable me to find a technology-based solution. After an extensive search, I came upon a small company that made a brain wave sensor and software module "games" that exercised the different cognitive centers of the brain, including memory. In fact, the company's tagline was "We can read your mind". Fascinated by this, I contacted the President and founder of the company and convinced him to join my wife and me over dinner so that he could assess her situation himself. After that, we went to my lab where he introduced both my wife and me to the company's product and learned how to use it in her recovery. For the next 8 months, we diligently trained on the software with the brain wave sensor. At the end of that period, my wife had remarkably regained 95% of her short-term memory. This made me wonder if other more sophisticated engineering tools and techniques could be brought to bare to improve our current understanding of human cognition. My career as an Aeronautics Professor at the University of Maryland had allowed me to use "Black Box" system Identification methods to model and analyse several exotic aerodynamics systems. My later arrival at the Texas A&M University gave me the freedom and resources to apply these techniques in an effort to identify a State Space cognitive model using bio-markers such as those garnered from Electroencephalograms (EEG) which measure the electrical activity of the brain. During the course of the research, I and my co-authors augmented the resulting model to handle weak nonlinearities and uncertainty using adaptive observers. Our models have informed us of Brain Wave Modes and given us a window into what we call "The Music of the Mind".

The science of musical sound has long been a human endeavor. The ancient Greeks have since the sixth century viewed mathematics and music as expressions of the harmony of nature. A string plucked generates a pitch. If a guitarist now only allows half the length of the same string to vibrate, the sound is a perfect octave higher. If two-thirds of the string may vibrate, the sound is a fifth higher. This remarkable phenomenon of whole number relationships between length and pitch is concretely harmonic and forms the basis of all music.

While instruments and the human voice can generate an infinite number of musical tones, each one can be described mathematically in terms of intensity, frequency, and waveform. The waveform, also called timbre, is especially important as it quantifies what makes a C note on a guitar different from the C on a violin. This effect is particularly influenced by the top plate of a stringed instrument. As a violin string is excited, it transmits vibrations to the top plate of the violin, which has natural resonant modes of vibration based on the shape and material of the top plate. Different strings excite many different resonant mode shapes in the top plate simultaneously. The same excitation results in different vibration mode shapes depending on the instrument. The sound of an instrument we hear is the sum total of all the modes excited by a string carried through the air. Understanding how the wood, fasteners, and shape of the instrument influence the timbre is a critical application of acoustic engineering. Master craftsmen create instruments that resonate at well separated frequencies, spanning the playing range of the instrument.

This book develops an analytical modeling technique to understand the "timbre" of brain waves. Brain waves are known to have harmonic vibratory patterns, but they are more complicated than the standing transverse waves of sound. This complicates the analysis and understanding of brain waves. While sound waves are measured with a microphone, brain waves are measured with electrodes. These electrodes capture the mass ensemble activity of neurons, because it is difficult to measure the oscillation of individual neurons in the same way as to measure the vibration of individual air molecules. Yet, the ensemble behavior of brain waves remains organized and vibratory.

The modeling technique herein was originally developed for the dynamic analysis of civil structures, especially as they were excited by earthquakes. Instruments are excited by precise excitations from musicians, but the excitation to the brain at a given moment is unclear, so the model must account for an unknown input. At its core, the method decomposes brain wave recordings into distinct vibratory modes, each with a given frequency.

This work develops the natural analogy between the modal view of instrument timbre and the modal view of brain wave "timbre". Each of the brain wave modes describes the spatial behavior of the brain wave at a specific frequency. Extending the musical analogy, each of the modes contributes individually to the music of the brain. Unlike stringed instruments, these waves are traveling in three dimensions across the brain and do not always move in phase with one another. Added together through a weighted superposition, the modes recreated the measured brain wave.

College Station, USA   James Hubbard Jr.

Contents

1 Introduction 1
   1.1 Cognition as a Dynamic Process 1
   1.2 Motivation: A Holistic View of Brain Waves 2
      1.2.1 What is a "model" of Brain Waves? 2
      1.2.2 Important Modeling Considerations for Brain Wave
           Dynamics 5
      1.2.3 Brain Wave Dynamics are Relevant to Many Modeling
           Outcomes 8
   1.3 Literature Review on Dynamic Models of Brain Waves 11
      1.3.1 Historical Developments and Early Approaches 11
      1.3.2 Limitations of Previous Approaches 14
   1.4 Proposed Approach 15
   1.5 Research Hypothesis 17
   1.6 Contributions of This Work 18
   1.7 Document Outline 19
   References 19

2 A Dynamic Systems View of Brain Waves 29
   2.1 Electroencephalography as a Cognitive Biomarker 29
      2.1.1 Relevant Characteristics of EEG for Dynamic Brain Wave
           Modeling 29
      2.1.2 Statistical Properties of EEG Measures 31
      2.1.3 Biological Sources of Corrupting Noise 32
      2.1.4 Inorganic Sources of Corrupting Noise 34
   2.2 A Canonical Approach to the Analysis of Brain Wave Dynamics 35
      2.2.1 Treating the Nonlinear Effects of Brain Waves 37
   2.3 Modal Analysis of State Space Brain Wave Models 38
      2.3.1 Modes Jointly Capture Space Time Dynamics 38
      2.3.2 Analytical Relevance of Eigenmodes 39
   2.4 System Identification Tools for Brain Wave Analysis 41
   References 42

3 System Identification of Brain Wave Modes Using EEG 45
  3.1 Introduction and Motivation 45
    3.1.1 Motivation: Dynamical Models of Biomarkers 45
    3.1.2 Linearization of Neural Dynamics 47
    3.1.3 Evaluation of System Identification Techniques for Brain
          Wave Modeling 47
    3.1.4 Overview of Considered Output-Only Algorithms 48
  3.2 Evaluation of Output-Only System Identification Techniques 52
    3.2.1 Model Variance Among Subjects 53
    3.2.2 Databases for Initial Brain Wave Modeling 53
    3.2.3 Joint Distributions of Modal Parameters 54
    3.2.4 Assumptions and Constraints 54
  3.3 Results 56
    3.3.1 Reducing the Number of EEG Channels 59
  3.4 Conclusions 60
    3.4.1 Recommendations 61
  References 61

4 Modal Analysis of Brain Wave Dynamics 65
  4.1 Research and Modeling Goals 65
  4.2 Technical Approach to Cognitive Modeling 66
    4.2.1 Adaptation of System Identification Algorithms for EEG Data ... 66
    4.2.2 Analysis of EEG Eigenmodes 67
    4.2.3 The Existence of Stimuli Independent Common Modes 71
  4.3 Results of a Subject Classification Task 72
    4.3.1 Experimental Validation and Verification Using a Neural
          Network Classifier 72
    4.3.2 An Extension to the EEG Motor Movement/Imagery Dataset 73
    4.3.3 Comparing OMA and DMD for Subject Identification 74
    4.3.4 Optimal System Representations and Neural Network
          Interpretation 75
  4.4 Conclusions 77
    4.4.1 Recommendations 77
  References 78

5 Adaptive Unknown Input Estimators 81
  5.1 Introduction 81
  5.2 Unknown Input Dynamics 84
  5.3 Input Generators as a Model of the Unknown Input 85
  5.4 Main Result: Adaptive Control Architecture for Unknown Input
    Estimation 87
    5.4.1 Composite Error Dynamics 88
    5.4.2 Proof of Composite Error Convergence 88

5.5   Illustrative Examples 90
5.6   Conclusions 94
References 94

6 Reconstructing the Brain Wave Unknown Input 97
6.1   Introduction and Motivation 97
6.2   Technical Approach 98
     6.2.1   Treating the Nonlinear Effects of Brain Waves 99
     6.2.2   Treating the Unknown Input 100
     6.2.3   Estimator Architecture and Proof of Convergence 101
     6.2.4   Datasets 107
6.3   Results 108
     6.3.1   Performance Benefits of UIO 109
     6.3.2   Analytical Benefits of UIO 110
     6.3.3   The Predictive Capability of the UIO 114
     6.3.4   Limitations of the Input Estimate 119
6.4   Conclusions 123
     6.4.1   Recommendations 124
References 124

7 Conclusions and Future Work 127
7.1   Summary 127
7.2   Key Contributions of This Work 128
     7.2.1   Modal Identification of Linear Brain Wave Dynamics 128
     7.2.2   Analysis of Spatio-Temporal Brain Wave Modes 129
     7.2.3   Theoretical Considerations for the Estimation of Unknown
            Inputs 129
     7.2.4   Online Estimation of Nonlinear Brain Wave Dynamics 129
7.3   Recommendations for Future Work 130
     7.3.1   Considering Multiple Data Types 130
     7.3.2   Improved Diagnostics and Analysis 130
     7.3.3   Spatial Filtering of Biomarkers 131
     7.3.4   Probabilistic Considerations 131
Reference 132

Introduction

1

1.1 Cognition as a Dynamic Process

This chapter motivates the topic of this work. Human consciousness, especially in the form of cognition and decision-making, is the single greatest input to the global economy and remains mostly unquantified. While we have used our brains to name, quantify, and manipulate the world around us, we have relatively little quantification of the brain itself [1]. Brain physiology, especially in the form of electrical and chemical processes, gives rise to observable, dynamic signals which may be correlated with human cognition and decision-making. In particular, synchronized electrical pulses of the brain's neurons yield a set of wave-like patterns known as *brain waves*. Brain waves are evidence of ensemble communication from one mass of neurons to another and exhibit spatio-temporal dynamics.

Correlations between brain waves and human behavior exist, including emotion [2], decision-making [3], and situational awareness [4]. However, nonlinear effects and inter-individual differences make deterministic modeling and analysis of these non-stationary brain waves difficult [5]. As a result, few existing techniques have translated into practical clinical and human performance applications. This uncertain dynamic process, which is the relationship between brain wave activity and human behavior, is defined for the purposes of this work as *cognition*. Accordingly, we refer to the human behavior outcomes of this process as *cognitive outcomes* throughout this work.

Coarsely, cognition, as defined here, can be viewed as a system that gives rise to mass neuronal activity in the brain as a result of conscious thought (Fig. 1.1). From this figure, one can imagine a very complicated "transfer function" describing the dynamics of brain waves and how they affect human behavior. Of course, it would be difficult to derive dynamical descriptions of brain wave activity from the first principles. However, there are a variety of engineering analysis tools for the identification and analysis of complicated, uncertain dynamical systems. There is an opportunity, explored in this work, to develop modern system

© The Author(s), under exclusive license to Springer Nature Switzerland AG 2023   1
T. D. Griffith et al., *A Modal Approach to the Space-Time Dynamics of Cognitive Biomarkers*, Synthesis Lectures on Biomedical Engineering, https://doi.org/10.1007/978-3-031-23529-0_1

Fig. 1.1 Brain wave dynamics as a black box

identification techniques for brain waves. The primary objective of this work is to develop spatio-temporal dynamical mappings of brain waves as an indicator of cognitive dynamics. Such mappings are called *models*, especially when they contain spatio-temporal dynamical relationships.

Improved models of brain wave dynamics offer an increased understanding of how human physiology influences cognition and decision-making with an eye toward safer human-robotic teams, sharper clinical diagnostics, and richer brain-computer interfaces. Here, a survey of existing work relating to modeling brain waves is taken. Following the survey, a research hypothesis is proposed for this work. The chapter closes with a discussion of the novel contributions in this work along with a summary of the complete document.

1.2 Motivation: A Holistic View of Brain Waves

In the past 50 years, new imaging and sensing technologies have generated a vibrant multi-disciplinary research community dedicated to imaging brain waves at both the micro and macroscopic levels. Yet, rigorous explanations for many ubiquitous cognitive outcomes remain elusive. For example, operator workload is a topic of significant research across safety-critical industries, but subjective surveys repeatedly outperform objective measures from physiological data streams [6–8]. Inter-individual differences, nonlinear effects, and data quality issues can easily spurn ad-hoc or piece-wise approaches [9]. Successful holistic modeling of brain wave dynamics may offer new insights into cognition and decision-making.

1.2.1 What is a "model" of Brain Waves?

Because of the difficulties associated with modeling the dynamics of cognition thus far, it is important to be specific about the potential of brain wave models. Recreating one-to-one models of the brain digitally remains difficult [10], and is *not* the goal of this work. It is, however, computationally feasible to identify models which match a given physiological data stream from brain waves. The ultimate goal of this process is to perform predictions on an individual basis of cognitive outcomes for clinical and human performance applications [11]. Recreating measured brain waves within established mathematical frameworks offers insight into the dynamics of brain wave activity, without the need for identical simulation

[12]. Modeling, as considered for the purposes of this work, consists of the synthesization of descriptions for dynamic physiological signals, within an established mathematical framework, for diagnosis, visualization, and analysis of brain waves.

1.2.1.1 Biomarkers are Indirect Measures of Brain Wave Dynamics

Brain waves can be measured *indirectly* with a variety of physiological data streams. Physiological data streams are often called biological markers (biomarkers). Biomarkers are "a characteristic that is objectively measured and evaluated as an indicator of normal biological processes, pathogenic processes, or pharmacologic responses to a therapeutic intervention" [13]. Biomarkers, as considered in this work, consist of a subset of these objective measurements that have spatio-temporal dynamics and may, therefore, be relevant to the spatio-temporal dynamics of brain waves. That is, this effort is focused on *cognitive biomarkers* [14] whose behavior can be considered an observable system output of brain waves.

The vast majority of brain wave models in the existing literature make use of noninvasive cognitive biomarkers for their studies [11]. Invasive measures, such as electrocorticography, have excellent signal to noise ratio and are a promising therapeutic for various cognitive ailments (e.g. epilepsy, paralysis), but can be difficult to implement in a study since they involve surgical implantation [15]. Accordingly, while there are studies analyzing brain waves with invasive cognitive biomarkers [16], they often have relatively few subjects in the study. There is a tradeoff between data quality and data quantity, which can make drawing group level conclusions difficult when using invasive cognitive biomarkers [17]. As a result, most researchers study many subjects with noninvasive cognitive biomarkers, rather than a few subjects with invasive biomarkers.

Noninvasive biomarkers include structural magnetic resonance imaging (sMRI), functional magnetic resonance imaging (fMRI), electroencephalography (EEG), magnetoencephalography (MEG), and diffusion tensor imaging (DTI). The majority of existing studies consider fMRI imaging as the primary cognitive biomarker [18]. fMRI is typically used over sMRI because fMRI detects hemodynamic responses through changes in blood oxygenation level-dependent (BOLD) signals and is, therefore, better suited to the correlation between tasks and regional changes in brain waves as seen in Fig. 1.2. MRI measures have greater spatial resolution than other noninvasive methods, so it is natural to study cognitive biomarkers, whose spatial dependence is well established [19], with MRI measures. However, fMRI has a limited temporal frequency, often <1 Hz, because it is reliant on a hemodynamic response [21]. While hemodynamics respond to neuronal activity, they are not a direct measure of neural oscillations [22]. Further, current MRI machines are difficult to integrate with many experiments because of their size and cost [23]. As a result, scalp EEG measures, which have millisecond temporal resolution and may be readily implemented in experiments, are a popular alternative to fMRI measures [24]. However, the increased temporal resolution comes at the expense of significantly decreased spatial resolution. State-of-the-art "high" density EEG devices have fewer than 400 spatial channels, which pales in comparison to the

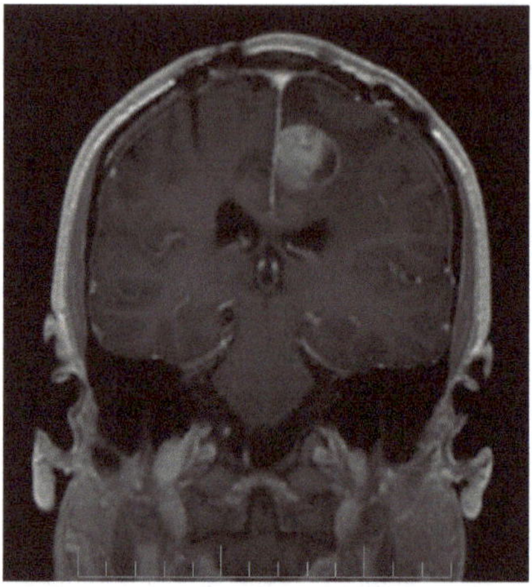
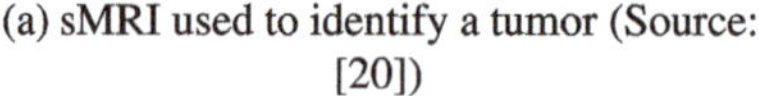

(a) sMRI used to identify a tumor (Source: [20])

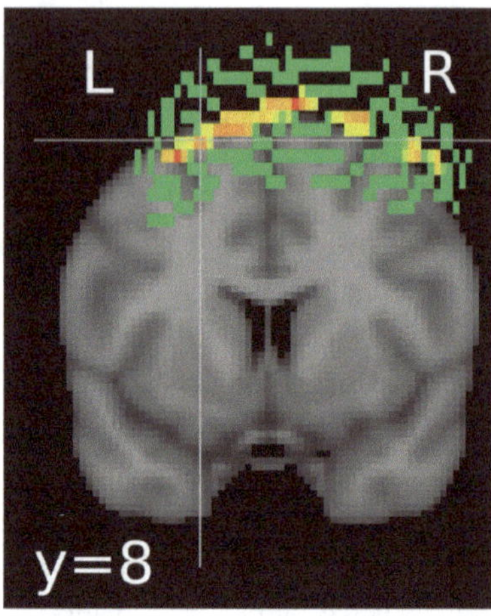

(b) fMRI used to identify hemodynamic changes

Fig. 1.2 Comparison between sMRI and fMRI. fMRI is well suited to measuring dynamic spatial responses from brain wave activity

$\approx$ 5-micron spatial resolution of fMRI measures. Within the last decade, electromagnetic shielding and sensing technologies of both fMRI and EEG devices have improved enough for both measures to be collected simultaneously without interference [25]. Much effort has been dedicated to jointly analyzing the spatial information in BOLD fMRI measures and the temporal information in EEG measures [26, 27]. There remains an unsolved problem to develop approaches that not only treat the cognitive outcome of interest, but also provide information on the dynamic relationship between the two measures since brain waves influence blood flow in the brain greatly. Multiple biomarker approaches are increasingly seen as important in the analysis of brain wave dynamics [28].

While fMRI and EEG measures are partially dependent sources of information [29], they do originate from different physiological changes in the brain. Conversely, MEG responses come from the same physiological activity as EEG responses. Even though MEG is theoretically not sensitive to radial components of dipolar sources while EEG is [30], the localization accuracy of the MEG and EEG has been directly compared and found to be of the same magnitude [31]. As a result, there is less motivation to jointly consider these two measures. EEG devices were invented and popularized before MEG devices, therefore EEG brain wave analysis tends to be more ubiquitous.

Finally, while DTI measures are seen as increasingly novel, the exact experimental protocols to generate valid data are evolving [32]. DTI is an MRI-derived measure that relies on changes in water diffusion to analyze the spatio-temporal dynamics of brain waves. While DTI measures are independent of fMRI, MEG, and EEG measures, the measure is still gaining popularity and there are fewer studies considering DTI [33].

The joint consideration of multiple data streams at different temporal and spatial scales is of great interest to many engineering applications, including brain-computer interfaces (BCIs) [34], autonomous robots [35], and structural health monitors [36]. Since cognitive biomarkers are not wholly independent sources of information, there is an opportunity for joint analysis of multiple biomarkers. Much progress has been made in identifying and eliminating contaminating sources of noise during simultaneous data acquisition (e.g. simultaneous fMRI-EEG). This opens opportunities for direct comparison and synthesization of simultaneous cognitive biomarker data streams.

In contrast to traditional engineering dynamics, the nature of the observable outputs of the cognitive system and their underlying source is unclear. At present, robust connections between the underlying cellular processes and common noninvasive biomarkers are unclear, including EEG [37], fMRI [38], and MEG [39]. However, these biomarkers are *correlated* with the underlying biology and their modeling and interpretation are useful in both clinical [40] and human performance applications [41]. This introduces significant complexity to the problem of dynamical modeling. Dynamics engineers are typically able to choose or modify the sensing suite for a given system to preserve desirable properties, such as observability. In studying brain waves, the biomarker selection is limited and they are not direct measures of brain waves. Curiously, [42] has postulated that the biomarkers may be encrypted, which suggests a proper transformation of physiological data may reveal much about the inner workings of the brain.

1.2.2   Important Modeling Considerations for Brain Wave Dynamics

The complexity of brain waves in time and space yields many modeling opportunities. It is important now to understand some of the high-level modeling considerations that are particularly relevant to brain wave dynamics, such as:

1. **Local versus Whole-Brain Dynamics:** It is well established that different spatial regions of the brain contribute to different biological tasks. Electrical signals from photoreceptors primarily feed the occipital lobe in the rear of the skull, while the olfactory epithelium sends scent signals to the olfactory bulb at the base of the forebrain. If the purpose of a given model is to describe the flow of information from photoreceptors to the brain, a local "searchlight" model of the occipital lobe may be sufficient. Alternatively, understanding how visual stimuli influence emotions is likely to require a model that incorporates spatial dynamics from the different regions of the brain, as one region may process the visual stimuli while another is responsible for generating an emotional response. While much of the original literature modeling biomarkers are concerned with predicting the local behavior of brain waves [43–45], it is increasingly clear that a whole-brain view, which accounts for the spatial dynamics as brain waves flow from one part of the brain to another, is needed to move from analysis to diagnosis [46, 47].

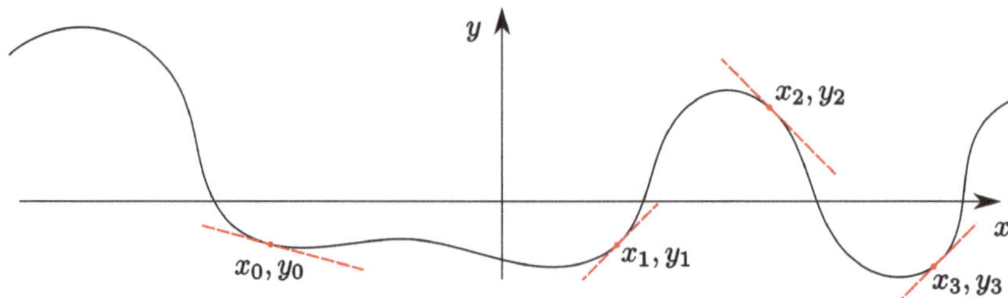

Fig. 1.3 A nonlinear function $y = f(x)$ characterized by four linear models. The number of linear models needed depends on how strong the nonlinearity is and how accurate the models must be

2. **Linear versus Nonlinear Models:** Almost all systems exhibit nonlinear behavior. Much of modern estimation theory is dedicated to analyzing and stabilizing nonlinear systems. Brain waves exhibit nonlinear effects, and these effects are important to modeling outcomes [48, 49]. Accordingly, much existing work has treated these nonlinear effects explicitly through parametrization [50–52]. Rather than considering the nonlinearity explicitly, it is also valid to linearize the observed dynamics for different operating neighborhoods. There are many examples of the use of multiple linearized models at different neighborhoods to describe nonlinear plant dynamics [53–55]. Consider the nonlinear function $f(x)$ in Fig. 1.3, which may represent the observed output y of a single channel of EEG data. The nonlinear function can be linearized using a Taylor series expansion around an operating point x_i as $f(x) = f(x_i) + \frac{df}{dx}|_{x=x_i}(x - x_i) + H.O.T.$ Each of the four linearizations has a different parametrization and is only valid over a certain range of x values. The linearization is as important as the model. The operating point x_i, y_i informs the state of the dynamical system in addition to the extracted dynamics. Each linearization could be a different transfer function or state space model for an observed cognitive biomarker. Whether the nonlinearities are explicitly treated or are treated with a collection of linear models, this behavior must be accounted for.

3. **Individual versus Population Models:** In order to draw scientific conclusions on brain wave activity, researchers often look at trends over numerous individuals [56]. Many studies apply their analysis over multiple subjects, resulting in a population view of brain waves [57, 58]. These averaged dynamics have revealed functional and structural insight, but exact relationships to cognition remain obscure [59]. Further, [60] reports that models are often more closely correlated with individuals than with the task the individuals perform, which supports the fact that even brain anatomy, and therefore, brain waves, varies from person to person [61]. Although the need for individualized models is well established [62, 63], translation to clinical application has been challenging as discussed in Sect. 1.2.3. There is an opportunity for individualized models to confirm group level conclusions.

4. **Static versus Dynamic Models:** Treating the process by which physiological signals give rise to observable biomarkers as dynamic is a somewhat new idea [64] and was considered static historically [65]. There is especially little knowledge of dynamic large-scale interactions between the spatial dynamics of brain waves and their cognitive outcomes [66]. In the same way that population level modeling reveals *some* insight into the population level activity of brain waves, static models provide a rigorous modeling framework and have yielded a wealth of baseline brain wave knowledge [67], primarily considering the identification and analysis of resting state networks (RSN) [57, 68]. Increasingly, the need to account for brain wave dynamics, in both space and time, is becoming clear [69] and accessible [70]. Of course, the consideration of dynamics introduces complexity to the modeling procedure as one must distinguish between temporal variability from noise sources and that which is relevant to cognitive outcomes. However, dynamic brain wave models have already demonstrated increased efficacy in the modeling of PTSD [71], schizophrenia [72], and vigilance [73].

5. **Offline versus Online Models:** The use of real-time brain wave modeling is typically driven by the mathematical framework of the selected model and the model's desired application. Offline models are those which cannot execute their modeling procedure simultaneously with the collection of the data stream. In modeling brain waves, offline models are particularly oriented for scientific outcomes, where the complexity of the hypothesis considered may require more intensive modeling and statistical considerations. For example, an offline approach is appropriate for modeling the effects of LSD on the brain [74], understanding how genomics influence cognitive disorders [75], and exploring the ability to identify distinct brain wave "states" [76]. In each of these cases, the need for computational complexity outweighs any need for online modeling. Conversely, models oriented toward individualized cognitive models are most useful in the context of Brain- Computer Interfaces (BCI), where online processing is needed for successful application [77]. One can imagine that a model tracking a pilot's engagement from physiological data streams is only useful if it runs in real time and can provide feedback to the crew. While it is non-trivial to simply collect physiological data in real time [78], improvements in sensing and computing show promise for truly real-time BCIs that support useful modeling outcomes such as instant imaging [79] and robot manipulation [80].

6. **Task Dependent versus Task Independent Models:** It has been established that biomarkers are more correlated with individuals than with tasks [60]. For example, it is much easier to distinguish the EEG signals of two different people at rest than it is to distinguish the EEG signal of a person at rest from the EEG signal of the same person performing a difficult mental task. As a result, task-dependent measures introduce significant complexity to the modeling process, since the response to the task must be separated from the inherent inter-individual difference. However, task-dependent measures may be tailored for a specific clinical or human performance outcome, often through the use of individualized modeling procedures. This tradeoff between biomarkers that are task

dependent (fMRI, EEG, and MEG) and those that are task independent (sMRI, DTI, and rest state EEG) must be considered early in the development of a given dynamical brain wave model. Task independent models in general tend to have simpler outcomes, but are accordingly more robust [81, 82].

7. **Parametric versus Nonparametric Models:** The parametrization of a model is less binary than the other properties in the discussion above. When describing the dynamics of a given system, some assumptions about the system are necessary to synthesize a rigorous model from the observed outputs. In many mechanical systems, it is possible to derive a general form of the dynamics from first principles, and then identify the specific *parameters* for the general form. In the simplest case, if a system is observed to have quadratic behavior, one would attempt to identify the parameters (a, b, c) in the general expression $ax^2 + bx + c = 0$. The model is *parametrized* as a quadratic. Alternatively, there may be no easily identified general form of the dynamics, in which case the modeling procedure is nonparametric. It would be difficult to derive the governing equations of brain waves from first principles, so some structure is generally imposed on the system. Since it is challenging to derive any type of equation for brain waves [83], the cognitive process is a black box. Knowledge of the system ends at the outputs. The box is completely opaque, and measurements of the outputs (which are only correlated to the underlying dynamics) are the only information about the system available. There are examples of completely nonparametric modeling [84], but even the use of a state space [85] or stochastic [86] framework offers some structure to the form of the brain wave dynamics.

Of these properties, there is a particular need for models to be *dynamic*, incorporating spatio-temporal variability at multiple scales [87]. While the uncertainty and nonlinearity associated with brain waves encourage static analysis, this can lead to problematic statistical analysis [88]. New models should above all else leverage rigorous analytical techniques to jointly capture the spatio-temporal behavior of brain waves as they traverse across the entire brain.

1.2.3 Brain Wave Dynamics are Relevant to Many Modeling Outcomes

Predicting the internal dynamics of brain waves is a novel engineering task. The field of dynamics has historically been oriented toward an application. Turbine dynamics inform design constraints, structural dynamics inform failure modes, and fluid dynamics inform rocket performance characteristics. The power of dynamics and estimation is the ability to predict the future, if only within a set of established assumptions. Similarly, the modeling of brain waves is most useful when developed with some modeling outcome in mind. Because human cognition is an input to so many systems, the set of possible modeling outcomes for dynamical brain wave models is particularly diverse. There is a distinction between two major types of dynamic brain wave models, those for clinical applications and those for human performance applications. In order to understand the field of brain wave modeling, a review of these applications is taken.

1.2.3.1 Models for Clinical Application

The most evident application for dynamic models of brain wave activity is in clinical applications. Successful modeling of a system's dynamics typically enables engineers to predict the system's behavior by revealing the internal structure of the system. While much progress has been made in basic neuroscience to understand and image the functional and structural details of the brain [89], translational neuroscientists have struggled to convert this knowledge into tools for basic clinical applications [90]. Historically, these tools are difficult to deploy and scale due to the inter individualization of each person's brain waves and the homogeneity of many research datasets [18, 91].

Driven by a recent wealth of predictive algorithms and computational capability, much promising work has been published on the modeling and imaging of brain wave data to predict numerous clinical outcomes. The vast majority of publications apply dynamic brain wave models for diagnosis and prognosis of clinically relevant outcomes. These include predictive models for Alzheimer's [92], psychosis [93], depression [94], autism [95], ADHD [96], Parkinson's [97], and PTSD [98].

Increasingly, models rely on machine learning or general pattern recognition algorithms to achieve an acceptable classification accuracy. This introduces a number of issues that hamper the widespread application of these models. First, while accuracy is the standard metric to evaluate the performance of a predictive machine learning model, it encourages group level modeling. This is naturally detrimental to the application of the model; whose performance is only valuable to the extent that it can predict outcomes for *individuals*. More importantly, [18] points out that many of the reported diagnostic accuracies exceed the reliability of the existing diagnoses from physician to physician. That is, the algorithm is more accurate than the humans who generated the labels initially. This conflict results from optimistic classification bias as a result of machine learning models which are overfitting. Dynamic brain wave models need to be *robust* and *individualized* for successful deployment.

The impact of successful diagnostics on these outcomes is difficult to overstate. Many cognitive dysfunctions are currently diagnosed by physicians on the basis of patient interviews. A physician would never perform an artery bypass based on how a patient felt about their heart plaque. The physician would get an echocardiogram and make decisions based on the resulting diagnostic test. Yet, many cognitive issues are diagnosed based on how patients "feel" (i.e. self report). Accordingly, models should have physical significance for interpretability. It is insufficient for a physician to record brain wave data, send it to an algorithm, and have the algorithm return only a categorical classification. In the same way, the echocardiogram reports the necessary parameters to diagnose artery blockages, so too should dynamic brain wave models *aid* the diagnosis of cognitive outcomes. Therefore, models should be tied to measures correlated with *physiology* (e.g. EEG, fMRI), be informed by neuroscientific principles, and be *human interpretable*.

1.2.3.2 Models for Human Performance

As the computing outcomes become more advanced, the computer becomes more entangled with its human operator [99]. This is a significant paradigm shift from the computer as a tool for a given task (Fig. 1.4a) to the computer as a collaborator which aids the completion of a task (Fig. 1.4b). Anecdotally, a calculator and a surgical robot both involve computers, but only the latter has agency in the physical world. Often, the role of the human operator in a collaborative teaming task becomes oriented toward supervision and decision-making.

Simply by including these digital agents, in a task alongside a human, a bi-directional flow of information and shared decision-making process exists [100]. There is necessarily a translation from physiological signals in the nervous system to interpretation by a digital agent. Because digital agents have been introduced specifically to aid the human decision-making process in uncertain, dynamic environments, the human operator must know that the digital agent is reliable and benevolent for the partnership to work as intended. Toward this end, significant progress has been made on an agent's ability to sense and understand its environment [101–103]. This includes advances in both agent decision-making [104] and the transparency of the agent's decision-making process [105]. Conversely, less progress has been made toward encoding the human decision-making process for shared decision-making [106]. Improvements on the human decision-making side of the team are just as important as those on the digital side, but extracting models of human decision-making is more difficult than extracting models of agent decision-making. Human decision-making is often non-deterministic and seemingly irrational [107]. Despite the increased difficulty, a translation of physiological brain wave signals to a digitally interpretable representation

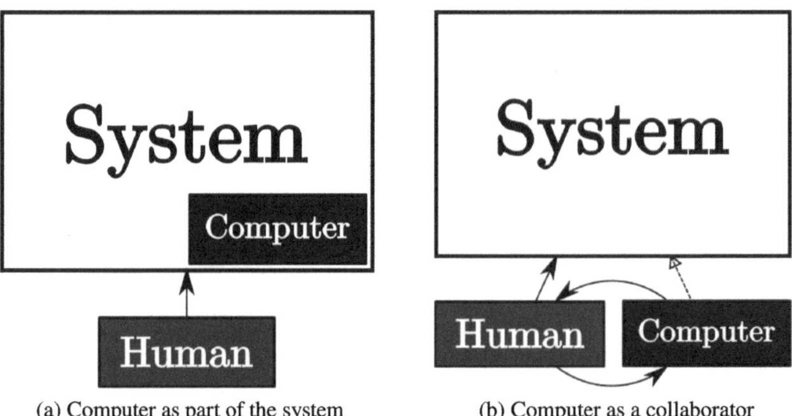

(a) Computer as part of the system   (b) Computer as a collaborator

Fig. 1.4 Two paradigms of digital electronics

of human cognition and decision-making remains necessary for effective teaming in the same way a translation of digital signals from an agent's sensors is necessary. A digital representation of cognition and decision-making gives the digital agent some representation of the human operator and enables shared understanding of the state of the team as feedback [108]. Therefore, rigorous models of human cognition and decision-making are needed for effective human-agent teaming.

1.3 Literature Review on Dynamic Models of Brain Waves

Having established the notion of brain wave dynamics and why they are important, this chapter turns to a review of state-of-the-art models of brain waves. As discussed in Sect. 1.2.3, there are a great many outcomes and so there are a great many models. Critically, while the promise of these models is encouraging, they have yet to make any meaningful impact on the public health sector. Therefore, the field is evolving, which makes it difficult to identify a "state-of-the-art" approach while the art itself remains largely unsolved. A quick search will reveal a multitude of modeling approaches making use of stochastics, nonlinear dynamics, artificial intelligence, and more. However, it is worthwhile to understand the historical developments of dynamical brain wave models, and where a new wealth of sensing and data sharing technologies are leading the field of translational neuroscience.

1.3.1 Historical Developments and Early Approaches

Today, it is well accepted that the brain is an electrochemical organ, but this was not always the case. Hans Berger, a German psychiatrist, was the first to record and publish results relating to the mass neuronal activity of brain waves in 1925. Berger left the work unpublished for five years for fear of controversy and skepticism [109], and it would be another four years before Berger's results were independently evaluated and then accepted: brains somehow generated wave-like electrical patterns that could be measured at the scalp [110]. Despite the slow initial reception, brain waves were quickly adopted for practical diagnostics of epilepsy once accepted [111]. These early results generated great enthusiasm for the study of brain waves.

Yet, it quickly became apparent that the temporal dynamics of brain waves were complicated because their statistical properties were non-stationary. While this made a time series view of brain waves undesirable, researchers soon realized that a spectral view of brain waves through the Fourier transform, made analysis and diagnosis significantly easier [112]. Even today, spectral decomposition and the Fourier transformation of cognitive biomarkers remains the most ubiquitous feature extraction technique for brain waves. From this early work, neuroscientists identified distinct brain wave frequency bands that correlated with different cognitive activities, ranging from the low frequency Delta waves (0.5–3 Hz) to the high frequency Gamma waves (32–100 Hz) [113].

By the 1960s, it became clear that neurons, specifically rhythmic, synchronized neuronal firings, were the source of brain waves [114]. Much work was done to evaluate the relationship between neuronal activity and brain wave recordings. Most crucially, [115], among others, realized that scalp brain wave recordings only capture electrical activity from neurons near the skull itself, which are now termed cortical neurons. Since this time, a great deal of effort has been dedicated to recording, decoding, and interpreting brain wave activity. Further, this era of brain wave study made clear that there is some voluntary control of brain wave activity [116]. That is, if one has a representation of their own brain waves, they exhibit some rudimentary control of the waves. This is a particularly encouraging fact for the application of engineering dynamics, where input-output relationships are critical. For a field that is less than one hundred years old, significant progress has been made.

Overall, the landscape of brain wave modeling, analysis, and diagnostics is evolving. Improvements in data sharing efforts have yielded a flood of high-quality biomarker data. This, in turn, is moving the community from single biomarker, task-independent group models of brain wave activity toward many biomarkers, individualized predictive models [14]. However, the list of brain wave models which are used in real clinical and human performance applications remains sparse. Predicting and monitoring Alzheimer's has been the most successful application of cognitive biomarker modeling so far [18]. Alzheimer's is the most common form of dementia, which is increasingly prevalent since many countries have aging populations. Historically, Alzheimer's is difficult to distinguish from natural cognitive decline without an autopsy [117], but recent advances in biomarker modeling suggest early diagnosis is possible. In turn, this should lead to early treatment and better clinical outcomes for affected patients.

The Spatial Pattern of Abnormality for Recognition of Early Alzheimer's Disease (SPARE-AD) is the state-of-the-art cognitive biomarker model at present [118]. SPARE-AD makes use of longitudinal sMRI scans to identify early indicators of brain atrophy, which indicates the onset of Alzheimer's. The modeling primarily consists of statistical analysis of anatomical maps in a stereotaxic space [119], which is shown to extract image features that are relevant to brain atrophy over a population. In some cases, these indicators may be determined before symptoms set in. The sMRI scans are evaluated by a machine learning model to predict cognitive decline. Overall, the process is a population-based, offline, task independent, statistical model, which is applying machine learning techniques to bridge the extracted features and the modeling outcomes. The dynamics here are primarily in the spatial domain, as subsequent scans occur on the order of months apart. This work is a hopeful example of how models of brain wave activity analysis may improve public health. Similar success has not been achieved with EEG-derived brain wave models, though they have received research attention [120].

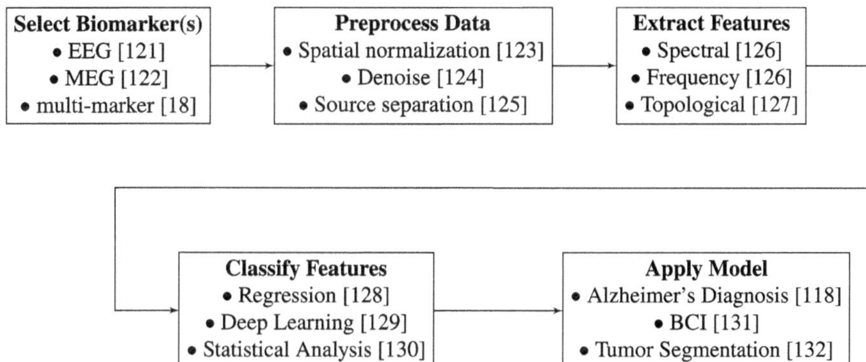

Fig. 1.5 A general architecture for dynamic brain wave modeling

1.3.1.1 Overview of Previous Approaches

The modeling specifics of each of the studies mentioned thus far are rather disparate. This reflects the often ad-hoc nature of cognitive modeling. In response, others have attempted to develop standard processing pipelines for EEG, fMRI, MEG, and other biomarkers [133, 134]. While any one of these has struggled to achieve widespread adoption, the research community agrees that more rigorous collection, processing, and analysis protocols are needed to improve the clinical and human performance applications of cognitive models [18].

Despite this, a general framework around most existing and developing dynamic brain wave models today is possible. Figure 1.5 identifies five major components of the most recent publications in brain wave modeling, with some of the most common methods under each heading. For example, increased accessibility and availability have made deep learning an increasingly popular method to extract useful information from extracted biomarker features. [129], accordingly, reviews the most up-to-date methods of deep learning classification for brain wave models.

As mentioned in Sect. 1.2.1.1, all dynamic brain wave models begin with the *selection of a biomarker* or set of biomarkers. More and more research points toward multi-marker approaches as a means of improving model performance [14]. Biomarker data streams, especially noninvasive imaging methods, have well established high signal to noise ratios. Therefore, most require some form of *preprocessing* before a useful set of features can be extracted from the raw data. In measures with high spatial density, for example, images need to be normalized to account for variations in skull and brain sizes. EEG signals are sensitive to electrical outlets, and so typically require filtering at the specific frequency of that country's power grid (e.g. 60 Hz in the United States). Note that these preprocessing steps can heavily influence or skew the resultant models [127]. With a *cleaner* representation of the data, any variety of transformations can be used to *extract features*, which are low dimensional representations of the data. EEG data is historically viewed with spectral features, while fMRI measures are often separated in terms of sparsity or independence. Extracted features are classically targeted at an application with some *classification method.*

1.3.2 Limitations of Previous Approaches

This review of the field has revealed the five key components of modern biomarker modeling. Advancements have been achieved in each of these, including biomarker data quality, preprocessing, feature extraction, classification, and targeted applications. As in any developing field, there are opportunities to extend the existing work, such as increased treatment of the following topics:

1. **Lack of generalizability across subjects, devices, and contexts** has limited the diagnostic usefulness of dynamic brain wave models. Historically, brain wave models have informed hypotheses on the functional behavior of brain waves. As these models have increasingly targeted clinical and human performance applications, the modeling protocols have maintained previous standards that were ideal for population level statistics. There is a need for models to work across populations on an individual level, especially for individuals not included in the initial development of the model. Just as important is the need for models to generalize across devices and laboratories. Biomarkers are measured with a wide variety of devices. EEG signals for example have multiple standards for sensor placement and channel count. New EEG models, therefore, should generalize across devices and be developed with robustness in mind. This includes the need for an increased model sharing effort among the research community. Increased efforts in sharing data, models, and analysis are already underway in the fields of deep learning, computer vision, and autonomy. Biological modelers have an opportunity to renew efforts in model availability and generalizability. For models to have a useful impact in their application, the research community as a whole must be able to survey and select the best approaches from a field of disparate methods. This demands access and repeatability.

2. **The joint spatio-temporal dynamics are often disregarded.** Recall that even temporal dynamics were only recently considered in models of brain wave dynamics. Further, many existing approaches are instantaneous measures that do not account for spatio-temporal dynamics jointly, even though these effects are well established in brain waves. New modeling approaches must be informed by theories of brain function for improved efficacy and insight. The brain has dynamic information flow through different regions at different times because of an anatomy that divides labor among different brain regions. Therefore, advanced models that treat this foundational property of the brain afford new opportunities to improve modeling outcomes for clinical applications and human performance.

3. **Approaches that seek to replace the human, rather than augment their decision-making** are difficult to interpret and translate. When models use human-generated labels

for classification (e.g. epilepsy diagnosis from EEG), it is important to remember the uncertainty associated with the human labels themselves. A model which reports $\geq$ 99% accuracy on labeled data has learned the biases and inaccuracies of those who labeled the data. This is a reason why many models have struggled to translate into clinical and human performance applications. There is no incentive to introduce a machine into the decision-making process if it makes the same mistakes as a human. Therefore, an argument is made supporting modeling approaches that select for low dimensional, human interpretable representations of brain wave dynamics, leaving flexibility for the human decision maker. For example, a model optimized for diagnosing epilepsy should reduce the EEG recordings to a limited set of interpretable features a doctor can use for diagnosis, rather than directly predicting the diagnosis with the model itself. Such an approach will generate models that describe the brain wave function, rather than simply fitting the desired outcome. Models of this form allow the computational part of the process to do what it does best (search and process through lots and lots of data), while simultaneously considering the strengths of the human decision maker (using past experience to make decisions in a new space).

1.4 Proposed Approach

The limitations in the previous section motivated the technical approach in this work. *In particular, there is a demonstrated need for analytically rigorous, spatio-temporal dynamic models of the human brain waves which generalize across devices and populations oriented toward clinical and human performance applications.* First, there is a need for *rigor* in the modeling approach. Engineering dynamics have treated a wide variety of systems with a rigorous, canonical approach termed system identification. Rather than try to build a dynamic model of brain wave activity from nothing, this approach leverages the existing body of knowledge which has successfully synthesized models for some very complicated systems. Most modern system identification approaches make use of gray or black box modeling approaches. A white box model can be derived entirely from the first principles. This is challenging in dynamic models of brain waves since governing equations for the physiological activity of the brain remain elusive and boundary conditions are unclear. Gray box approaches, which make use of some limited physical insight in combination with input-output matching, cover the vast majority of physical dynamical systems in use today. Again, there are no widely accepted parametric descriptions for brain wave dynamics. Therefore, for the purposes of this work, the cognitive process which generates brain waves must be treated as a *black box*. Black box modeling makes use of measurable system data streams in order to formulate some limited description of the system dynamics. Such approaches typically limit interpretability and physical significance. A significant portion of this work will explore the application and interpretation of black box models as they relate to brain waves.

The black box approach to brain wave dynamics is limited by a significant constraint: the input to the brain is unobservable. Consider, for instance, a subject viewing an image of a rose. While the image is well defined in source space (i.e. each pixel is encoded in binary), it is less clear how the image is defined in "brain wave space". Generally, the system input can be carefully controlled to excite important system dynamics for identification. A gust of wind generates strain on an airfoil which changes the resistance of a Wheatstone bridge which emits a digital signal representing the force of the gust. This relationship does not currently exist for brain waves. Accordingly, the approach here must make use of black box *output-only system identification* techniques. As mentioned in Sect. 1.2.1.1, even the available methods for measuring brain waves remain only somewhat coupled to the underlying brain wave. In this work, we propose an approach that makes use of generic time series biomarker data and is, therefore, largely agnostic to the source of the data. In turn, this allows the approach presented here to support single or multi-marker recordings.

Special attention is needed to capture the joint spatio-temporal dynamics of brain waves. Brain waves may be considered a distributed parameter system of the form

$$\frac{dx}{dt} = f(x, t; \mu), \tag{1.1}$$

where $x \in \mathbb{X}$ is the brain wave system state at time t given parameters μ which may represent control or disturbance inputs. $f(\cdot)$, which describes the brain wave dynamics, is typically coupled and nonlinear. Formulating a nonlinear $f(\cdot)$ from a set of measurements $y = g(x)$ remains an open problem in system identification. Here, neither the form of $f(\cdot)$, nor the boundary conditions of brain wave dynamics are known. In order to make the problem solvable, $f(\cdot)$ is approximated with a discretized system of coupled ordinary differential equations (ODEs), which represent the brain wave plant at a *local* linearity as

$$\dot{x}(t) = Ax(t). \tag{1.2}$$

The *global* behavior of brain waves may be described with a discrete set of ODEs at each local linearity. While this may seem dubious, EEG signals are known to be highly non-stationary and nonlinear. To account for this while maintaining model performance and computational tractability, multiple linear points are a robust approach. Crucially to the analysis of brain wave dynamics, the matrix A inherently admits modes. Each eigenvector v of A has an associated eigenvalue λ. Any $x(t)$, which is the state of the brain waves at a given time t, can be rewritten as

$$x(t) = \sum_{i=1}^{n} \lambda_i^t v_i w_i^T \hat{x}(0) \tag{1.3}$$

which is a linear combination of modes. Brain waves may be described as a weighted linear combination of modes around a given operating point. w_i are the left eigenvectors corresponding to λ_i. Equation (1.3) is referred to as the modal decomposition of the system.

At every time step, the total brain wave dynamics may be represented as a sum of modes. Each eigenvector v_i has a corresponding modal frequency λ_i. v_i is a vector describing the mode shape. Relative to the brain wave activity, v_i elegantly reveals how the brain waves flow from different spatial regions of the brain to another at different temporal frequencies. Together, v_i and λ_i form an eigenmode. The eigenvectors v_i describe the spatial relationship at each temporal frequency λ_i. It is the objective of this work to identify high-fidelity modes which capture the spatio-temporal activity of brain waves. The modes act as a basis for the observed dynamics, but they need not be orthogonal.

The resultant modal decomposition of brain wave activity yields a rigorous, canonical description of the observed measurements that elegantly captures the spatio-temporal dynamics in state space terms. The mode shapes represent the regional connectivity of the brain waves and are relevant to modeling outcomes. These decompositions reduce the complexity of brain wave recordings significantly and are viewed as a tool for the imaging and analysis of cognitive biomarkers.

Further, by leveraging the canonical state space form of Eq. (1.2), a wealth of optimal estimation theory is available to us. This includes unknown input estimation, which is used to quantify a "brain wave space" view of the system input. In combination, this modeling approach makes use of well accepted tools in system identification and engineering dynamics to formulate locally linear modal decompositions of biomarker data which naturally and elegantly show the spatio-temporal behavior of brain waves in a human interpretable way, along with an estimate of the input. This process works in near real time, and is applicable to the clinical and human performance applications discussed in Sect. 1.2.1.

1.5   Research Hypothesis

There is a need for canonical modeling approaches to the dynamics of cognitive biomarkers for feature extraction and classification. Brain waves are of particular importance to clinical and human performance applications. Many biomarker data streams are generally seen as having a low spatial resolution, but are the most prevalent measurements for brain waves that may be used online in an application. Although output-only system identification techniques are generally restricted to large engineering structures under unknown loads, they may be extended for the modeling of brain waves. This work addresses the following research hypothesis:

> Brain waves are correlated with cognitive processes and admit modal decomposition. Modal decomposition, through modern output-only system identification techniques, is appropriate for the extraction of high-fidelity spatio-temporal dynamical descriptions of brain wave states when the known nonlinear effects are treated. These human

interpretable, high-fidelity modes are a rigorous tool for the analysis of cognitive processes oriented toward clinical and human performance applications.

Here, the focus primarily remains on the modeling of brain waves, especially their spatial dynamics. Since the approach is oriented toward modeling cognitive biomarkers, only EEG data is treated, but the synthesization of Eq. (1.2) is agnostic to data modality and may be extended to a multi-marker approach. Applications are used to demonstrate the efficacy and potential of the modeling approach as illustrative examples.

1.6   Contributions of This Work

In this chapter, we have motivated the need for a rigorous, spatio-temporal view of brain wave dynamics to improve clinical and human performance cognitive outcomes. This work proposes the development of improved brain wave models, through the use of modern system identification techniques. Because the spatio-temporal dynamics are especially important here, the modeling process employs a modal view. Brain wave eigenmodes leverage the inherent structure of state space dynamical modeling for the diagnosis and analysis of brain waves. From these models, adaptive optimal estimation techniques estimate the unknown input to the brain waves, also in the modal view. The main contributions of the present work can be summarized as:

- Development of a novel approach for the modeling of cognitive processes using modern system identification techniques that accommodates multi-marker data;
- Modal identification of brain waves, which canonically and elegantly capture the joint spatio-temporal dynamics of mass neuronal activity;
- Development of a novel adaptive unknown input observer, which uses optimal state estimation to synthesize the brain wave input in real time; and
- Application and analysis of the adaptive unknown input observer to brain wave dynamics, which estimates the exogenous information to the brain wave plant while updating the linear eigenmodes in real time when nonlinear effects are present.

While the demonstrations presented in this work primarily focus on the analysis of EEG data streams, the modeling process generalizes to other biomarkers that may be represented with a state space formulation, such as magnetoencephalography, electrocorticography, and electrocardiography. More broadly, while the identification of systems with output-only modal analysis has generally been reserved for use in large engineering structures such as bridges or boat hulls, this work demonstrates the broad applicability of modern system identification techniques and optimal estimation in the analysis of complicated input-output processes.

1.7   Document Outline

The body of this document progresses with a detailed look at EEG measures and their implications for brain wave dynamical models in Chap. 2. This is followed by a discussion and evaluation of modal identification techniques for EEG brain wave dynamics in Chap. 3. Chapter 4 details a systematized modeling practice for the identification and analysis of brain wave dynamical models. Then, this work turns to treat the nonlinear effects and general uncertainty of brain waves by introducing a generic estimator architecture for the online estimation of nonlinear brain wave dynamics in Chap. 5. This estimator is then applied and its performance is discussed in the context of brain waves in Chap. 6, which demonstrates in its entirety the highly efficacious modeling approach developed here. Conclusions and recommendations follow in the closing chapter, Chap. 7.

References

1. R. Adolphs, The unsolved problems of neuroscience. Trends Cognit. Sci. **19**(4), 173–175 (2015). https://www.sciencedirect.com/science/article/pii/S1364661315000236
2. T.-M. Li, H.-C. Chao, J. Zhang, Emotion classification based on brain wave: a survey. HCIS **9**(1), 1–17 (2019)
3. P. Golnar-Nik, S. Farashi, M.-S. Safari, The application of eeg power for the prediction and interpretation of consumer decision-making: a neuromarketing study. Physiol. Behav. **207**, 90–98 (2019)
4. L.G. Yeo, H. Sun, Y. Liu, F. Trapsilawati, O. Sourina, C.-H. Chen, W. Mueller-Wittig, W.T. Ang, Mobile eeg-based situation awareness recognition for air traffic controllers, in *2017 IEEE International Conference on Systems, Man, and Cybernetics (SMC)* (IEEE, 2017), pp. 3030–3035
5. Y. Shen, G.B. Giannakis, B. Baingana, Nonlinear structural vector autoregressive models with application to directed brain networks. IEEE Trans. Signal Process. **67**(20), 5325–5339 (2019)
6. L. Lecoutre, S. Lini, C. Bey, Q. Lebour, P.-A. Favier, Evaluating eeg measures as a workload assessment in an operational video game setup, in *PhyCS* (2015), pp. 112–117
7. M. Endsley, Situation awareness global assessment technique (sagat), in *Proceedings of the IEEE 1988 National Aerospace and Electronics Conference*, vol. 3. (1988), pp. 789–795
8. L. Mazur, P. Mosaly, L. Hoyle, E. Jones, L. Marks, Subjective and objective quantification of physician's workload and performance during radiation therapy planning tasks. Pract. Rad. Oncol. **3**, e171-177 (2013)
9. R. Petrican, C. Saverino, R. Shayna Rosenbaum, C. Grady, Inter-individual differences in the experience of negative emotion predict variations in functional brain architecture. NeuroImage **123**, 80–88 (2015). https://www.sciencedirect.com/science/article/pii/S1053811915007430
10. S. Kunkel, T. Potjans, J. Eppler, H.E. Plesser, A. Morrison, M. Diesmann, Meeting the memory challenges of brain-scale network simulation. Front. Neuroinf. **5**, 35 (2012). https://www.frontiersin.org/article/10.3389/fninf.2011.00035
11. J.D. Gabrieli, S.S. Ghosh, S. Whitfield-Gabrieli, Prediction as a humanitarian and pragmatic contribution from human cognitive neuroscience. Neuron **85**(1), 11–26 (2015). https://www.sciencedirect.com/science/article/pii/S0896627314009672

12. K. Gramann, T.-P. Jung, D.P. Ferris, C.-T. Lin, S. Makeig, Toward a new cognitive neuroscience: modeling natural brain dynamics. Front. Hum. Neurosci. **8**, 444 (2014). https://www.frontiersin.org/article/10.3389/fnhum.2014.00444

13. A. Atkinson, W. Colburn, V. Degruttola, D. Demets, G. Downing, D. Hoth, J. Oates, C. Peck, R. Schooley, B. Spilker, J. Woodcock, S. Zeger, Biomarkers and surrogate endpoints: preferred definitions and conceptual framework*. Clin. Pharmacol. Ther. **69**, 89–95 (2001)

14. J. Sui, R. Jiang, J. Bustillo, V. Calhoun, Neuroimaging-based individualized prediction of cognition and behavior for mental disorders and health: methods and promises. Biolog. Psychiatry **88**(11), 818–828 (2020). Neuroimaging Biomarkers of Psychological Trauma. https://www.sciencedirect.com/science/article/pii/S0006322320301116

15. D. Rubin, *Clinical Neurophysiology*, Contemporary Neurology Series (Oxford University Press, 2021). https://books.google.com/books?id=qG8vEAAAQBAJ

16. Y. Shiraishi, Y. Kawahara, O. Yamashita, R. Fukuma, S. Yamamoto, Y. Saitoh, H. Kishima, T. Yanagisawa, Neural decoding of electrocorticographic signals using dynamic mode decomposition. J. Neural Eng. **17**(3), 036009 (2020). https://doi.org/10.1088/1741-2552/ab8910

17. A.D. Kaplan, Q. Cheng, P. Karande, E. Tran, M. Bijanzadeh, H. Dawes, E. Chang, Localization of emotional affect in electrocorticography using a model based discrimination measure, in *2019 53rd Asilomar Conference on Signals, Systems, and Computers.* (IEEE, 2019), pp. 1709–1713

18. C.-W. Woo, L.J. Chang, M.A. Lindquist, T.D. Wager, Building better biomarkers: brain models in translational neuroimaging. Nat. Neurosci. **20**(3), 365–377 (2017). https://doi.org/10.1038/nn.4478

19. B. Biswal, F. Zerrin Yetkin, V.M. Haughton, J.S. Hyde, Functional connectivity in the motor cortex of resting human brain using echo-planar mri. Magn. Reson. Med. **34**(4), 537–541 (1995). https://onlinelibrary.wiley.com/doi/abs/10.1002/mrm.1910340409

20. R. Mommaerts, *MRI COR 03232011* (2011). https://flic.kr/p/9xVbcu

21. M.J. Sturzbecher, D.B. de Araujo, Simultaneous eeg-fmri: integrating spatial and temporal resolution, in *The Relevance of the Time Domain to Neural Network Models* (Springer, 2012), pp. 199–217

22. K.J. Friston, Modalities, modes, and models in functional neuroimaging. Science **326**(5951), 399–403 (2009). https://www.science.org/doi/abs/10.1126/science.1174521

23. L.L. Wald, P.C. McDaniel, T. Witzel, J.P. Stockmann, C.Z. Cooley, Low-cost and portable mri. J. Magn. Reson. Imaging **52**(3), 686–696 (2020). https://onlinelibrary.wiley.com/doi/abs/10.1002/jmri.26942

24. H.-J. Hwang, S. Kim, S. Choi, C.-H. Im, Eeg-based brain-computer interfaces: a thorough literature survey. Int. J. Hum.-Comput. Inter. **29**(12), 814–826 (2013). https://doi.org/10.1080/10447318.2013.780869

25. R.J. Huster, S. Debener, T. Eichele, C.S. Herrmann, Methods for simultaneous eeg-fmri: an introductory review. J. Neurosci. **32**(18), 6053–6060 (2012)

26. M. Bullock, G.D. Jackson, D.F. Abbott, Artifact reduction in simultaneous eeg-fmri: a systematic review of methods and contemporary usage. Front. Neurol. **12**, 193 (2021). https://www.frontiersin.org/article/10.3389/fneur.2021.622719

27. Y. He, M. Steines, J. Sommer, H. Gebhardt, A. Nagels, G. Sammer, T. Kircher, B. Straube, Spatial-temporal dynamics of gesture-speech integration: a simultaneous eeg-fmri study. Brain Struct. Funct. **223**(7), 3073–3089 (2018)

28. M.A. Pisauro, E. Fouragnan, C. Retzler, M.G. Philiastides, Neural correlates of evidence accumulation during value-based decisions revealed via simultaneous eeg-fmri. Nat. Commun. **8**(1), 1–9 (2017)

29. M. Prestel, T.P. Steinfath, M. Tremmel, R. Stark, U. Ott, fmri bold correlates of eeg independent components: spatial correspondence with the default mode network. Front. Hum. Neurosci. **12**, 478 (2018). https://www.frontiersin.org/article/10.3389/fnhum.2018.00478

30. S.P. Ahlfors, J. Han, J.W. Belliveau, M.S. Hämäläinen, Sensitivity of meg and eeg to source orientation. Brain Topogr. **23**(3), 227–232 (2010)

31. D. Cohen, B.N. Cuffin, K. Yunokuchi, R. Maniewski, C. Purcell, G.R. Cosgrove, J. Ives, J.G. Kennedy, D.L. Schomer, Meg versus eeg localization test using implanted sources in the human brain. Ann. Neurol. **28**(6), 811–817 (1990). https://onlinelibrary.wiley.com/doi/abs/10.1002/ana.410280613

32. J. Soares, P. Marques, V. Alves, N. Sousa, A hitchhiker's guide to diffusion tensor imaging. Front. Neurosci. **7**, 31 (2013). https://www.frontiersin.org/article/10.3389/fnins.2013.00031

33. O.A. Zaninovich, M.J. Avila, M. Kay, J.L. Becker, R.J. Hurlbert, N.L. Martirosyan, The role of diffusion tensor imaging in the diagnosis, prognosis, and assessment of recovery and treatment of spinal cord injury: a systematic review. Neurosur. Focus FOC **46**(3), E7 (2019). https://thejns.org/focus/view/journals/neurosurg-focus/46/3/article-pE7.xml

34. H. Gürkök, A. Nijholt, Brain-computer interfaces for multimodal interaction: a survey and principles. Int. J. Hum.-Comput. Inter. **28**(5), 292–307 (2012). https://doi.org/10.1080/10447318.2011.582022

35. M. Person, M. Jensen, A.O. Smith, H. Gutierrez, Multimodal fusion object detection system for autonomous vehicles. J. Dynam. Syst. Meas. Control **141**(7), 071017 (2019). https://doi.org/10.1115/1.4043222

36. A. Nasrollahi, W. Deng, Z. Ma, P. Rizzo, Multimodal structural health monitoring based on active and passive sensing. Struct. Health Monit. **17**(2), 395–409 (2018). https://doi.org/10.1177/1475921717699375

37. P.A. Robinson, C.J. Rennie, D.L. Rowe, S.C. O'Connor, J.J. Wright, E. Gordon, R.W. Whitehouse, Neurophysical modeling of brain dynamics. Neuropsychopharmacology **28**(1), S74–S79 (2003). https://doi.org/10.1038/sj.npp.1300143

38. K. Uludağ, K. Uğurbil, *Physiology and Physics of the fMRI Signal* (Springer US, Boston, 2015), pp. 163–213. https://doi.org/10.1007/978-1-4899-7591-1_8

39. J.A. Kim, K.D. Davis, Magnetoencephalography: physics, techniques, and applications in the basic and clinical neurosciences. J. Neurophysiol. **125**(3), 938–956 (2021). pMID: 33567968. https://doi.org/10.1152/jn.00530.2020

40. J.A. Roberts, K.J. Friston, M. Breakspear, Clinical applications of stochastic dynamic models of the brain, part ii: a review. Biol. Psychiatry: Cognit. Neurosci. Neuroimag. **2**(3), 225–234 (2017). https://www.sciencedirect.com/science/article/pii/S2451902217300149

41. Z. He, Z. Li, F. Yang, L. Wang, J. Li, C. Zhou, J. Pan, Advances in multimodal emotion recognition based on brain–computer interfaces. Brain Sci. **10**(10) (2020). https://www.mdpi.com/2076-3425/10/10/687

42. E.L. Dyer, M. Gheshlaghi Azar, M.G. Perich, H.L. Fernandes, S. Naufel, L.E. Miller, K.P. Körding, A cryptography-based approach for movement decoding. Nat. Biomed. Eng. **1**(12), 967–976 (2017). https://doi.org/10.1038/s41551-017-0169-7

43. Y. Kamitani, F. Tong, Decoding the visual and subjective contents of the human brain. Nat. Neurosci. **8**(5), 679–685 (2005). https://doi.org/10.1038/nn1444

44. J.V. Haxby, M.I. Gobbini, M.L. Furey, A. Ishai, J.L. Schouten, P. Pietrini, Distributed and overlapping representations of faces and objects in ventral temporal cortex. Science **293**(5539), 2425–2430 (2001). https://science.sciencemag.org/content/293/5539/2425

45. N. Kriegeskorte, R. Goebel, P. Bandettini, Information-based functional brain mapping. Proc. Natl. Acad. Sci. **103**(10), 3863–3868 (2006). https://www.pnas.org/content/103/10/3863

46. T. Hahn, T. Kircher, B. Straube, H.-U. Wittchen, C. Konrad, A. Ströhle, A. Wittmann, B. Pfleiderer, A. Reif, V. Arolt, U. Lueken, Predicting treatment response to cognitive behavioral therapy in panic disorder with agoraphobia by integrating local neural information. JAMA Psychiatry **72**(1), 68–74 (2015). https://doi.org/10.1001/jamapsychiatry.2014.1741

47. L.J. Chang, P.J. Gianaros, S.B. Manuck, A. Krishnan, T.D. Wager, A sensitive and specific neural signature for picture-induced negative affect. PLOS Biol. **13**(6), 1–28 (2015). https://doi.org/10.1371/journal.pbio.1002180

48. D. Garrett, D. Peterson, C. Anderson, M. Thaut, Comparison of linear, nonlinear, and feature selection methods for eeg signal classification. IEEE Trans. Neural Syst. Rehabil. Eng. **11**(2), 141–144 (2003)

49. K. Hoemann, L.F. Barrett, Concepts dissolve artificial boundaries in the study of emotion and cognition, uniting body, brain, and mind. Cognit. Emot. **33**(1), 67–76 (2019). pMID: 30336722. https://doi.org/10.1080/02699931.2018.1535428

50. C. Büchel, R. Wise, C. Mummery, J.-B. Poline, K. Friston, Nonlinear regression in parametric activation studies. NeuroImage **4**(1), 60–66 (1996). https://www.sciencedirect.com/science/article/pii/S1053811996900294

51. V.A. Maksimenko, A.E. Hramov, V.V. Grubov, V.O. Nedaivozov, V.V. Makarov, A.N. Pisarchik, Nonlinear effect of biological feedback on brain attentional state. Nonlinear Dynam. **95**(3), 1923–1939 (2019). https://doi.org/10.1007/s11071-018-4668-1

52. P.S. Pal, R. Kar, D. Mandal, S.P. Ghoshal, Parametric identification with performance assessment of wiener systems using brain storm optimization algorithm. Circuits Syst. Signal Proc. **36**(8), 3143–3181 (2017). https://doi.org/10.1007/s00034-016-0464-7

53. J. Roubal, P. Husek, J. Stecha, Linearization: students forget the operating point. IEEE Trans. Educ. **53**(3), 413–418 (2010)

54. C.-T. Chen, B. Shafai, *Linear System Theory and Design*, vol. 3 (Oxford University Press, New York, 1999)

55. P. Antsaklis, A. Michel, *A Linear Systems Primer* (Birkhäuser, Boston, 2007). https://books.google.com/books?id=7W4Rbqw_8vYC

56. E.P. Torres, E.A. Torres, M. Hernández-Álvarez, S.G. Yoo, Eeg-based bci emotion recognition: a survey. Sensors **20**(18) (2020). https://www.mdpi.com/1424-8220/20/18/5083

57. J.S. Damoiseaux, S.A.R.B. Rombouts, F. Barkhof, P. Scheltens, C.J. Stam, S.M. Smith, C.F. Beckmann, Consistent resting-state networks across healthy subjects. Proc. Natl. Acad. Sci. **103**(37), 13 848–13 853 (2006). https://www.pnas.org/content/103/37/13848

58. B.T. Thomas Yeo, F.M. Krienen, J. Sepulcre, M.R. Sabuncu, D. Lashkari, M. Hollinshead, J.L. Roffman, J.W. Smoller, L. Zöllei, J.R. Polimeni, B. Fischl, H. Liu, R.L. Buckner, The organization of the human cerebral cortex estimated by intrinsic functional connectivity. J. Neurophysiol. **106**(3), 1125–1165 (2011). pMID: 21653723. https://doi.org/10.1152/jn.00338.2011

59. H.-J. Park, K. Friston, Structural and functional brain networks: from connections to cognition. Science **342**(6158) (2013). https://science.sciencemag.org/content/342/6158/1238411

60. M. Nentwich, L. Ai, J. Madsen, Q.K. Telesford, S. Haufe, M.P. Milham, L.C. Parra, Functional connectivity of eeg is subject-specific, associated with phenotype, and different from fmri. NeuroImage **218**, 117001 (2020). https://www.sciencedirect.com/science/article/pii/S1053811920304870

61. S.A. Valizadeh, F. Liem, S. Mérillat, J. Hänggi, L. Jäncke, Identification of individual subjects on the basis of their brain anatomical features. Sci. Rep. **8**(1), 5611 (2018). https://doi.org/10.1038/s41598-018-23696-6

62. V. Jirsa, T. Proix, D. Perdikis, M. Woodman, H. Wang, J. Gonzalez-Martinez, C. Bernard, C. Bénar, M. Guye, P. Chauvel, F. Bartolomei, The virtual epileptic patient: individualized

whole-brain models of epilepsy spread. NeuroImage **145**, 377–388 (2017). Individual Subject Prediction. https://www.sciencedirect.com/science/article/pii/S1053811916300891

63. E.S. Finn, X. Shen, D. Scheinost, M.D. Rosenberg, J. Huang, M.M. Chun, X. Papademetris, R.T. Constable, Functional connectome fingerprinting: identifying individuals using patterns of brain connectivity. Nat. Neurosci. **18**(11), 1664–1671 (2015). https://doi.org/10.1038/nn.4135

64. F. Esposito, A. Bertolino, T. Scarabino, V. Latorre, G. Blasi, T. Popolizio, G. Tedeschi, S. Cirillo, R. Goebel, F. Di Salle, Independent component model of the default-mode brain function: assessing the impact of active thinking. Brain Res. Bull. **70**(4), 263–269 (2006). https://www.sciencedirect.com/science/article/pii/S0361923006002073

65. K.J. Friston, Functional and effective connectivity: a review. Brain Connect. **1**(1), 13–36 (2011). pMID: 22432952. https://doi.org/10.1089/brain.2011.0008

66. M. Breakspear, Dynamic models of large-scale brain activity. Nat. Neurosci. **20**(3), 340–352 (2017). https://doi.org/10.1038/nn.4497

67. C.F. Beckmann, M. DeLuca, J.T. Devlin, S.M. Smith, Investigations into resting-state connectivity using independent component analysis. Philosoph. Trans. R. Soc. B: Biol. Sci. **360**(1457), 1001–1013 (2005). https://royalsocietypublishing.org/doi/abs/10.1098/rstb.2005.1634

68. J.D. Power, A.L. Cohen, S.M. Nelson, G.S. Wig, K.A. Barnes, J.A. Church, A.C. Vogel, T.O. Laumann, F.M. Miezin, B.L. Schlaggar, S.E. Petersen, Functional network organization of the human brain. Neuron **72**(4), 665–678 (2011). https://www.sciencedirect.com/science/article/pii/S0896627311007926

69. R.M. Hutchison, T. Womelsdorf, E.A. Allen, P.A. Bandettini, V.D. Calhoun, M. Corbetta, S. Della Penna, J.H. Duyn, G.H. Glover, J. Gonzalez-Castillo, D.A. Handwerker, S. Keilholz, V. Kiviniemi, D.A. Leopold, F. de Pasquale, O. Sporns, M. Walter, C. Chang, Dynamic functional connectivity: promise, issues, and interpretations. NeuroImage **80**, 360–378 (2013). Mapping the Connectome. https://www.sciencedirect.com/science/article/pii/S105381191300579X

70. M. Rabinovich, K. Friston, P. Varona, *Principles of Brain Dynamics: Global State Interactions*, Computational Neuroscience (MIT Press, 2012). https://books.google.com/books?id=KOWZ2sZNlWQC

71. C. Jin, H. Jia, P. Lanka, D. Rangaprakash, L. Li, T. Liu, X. Hu, G. Deshpande, Dynamic brain connectivity is a better predictor of ptsd than static connectivity. Hum. Brain Map. **38**(9), 4479–4496 (2017). https://onlinelibrary.wiley.com/doi/abs/10.1002/hbm.23676

72. Ü. Sakoğlu, G.D. Pearlson, K.A. Kiehl, Y.M. Wang, A.M. Michael, V.D. Calhoun, A method for evaluating dynamic functional network connectivity and task-modulation: application to schizophrenia. Magn. Reson. Mat. Phys. Biol. Med. **23**(5), 351–366 (2010). https://doi.org/10.1007/s10334-010-0197-8

73. G.J. Thompson, M.E. Magnuson, M.D. Merritt, H. Schwarb, W.-J. Pan, A. McKinley, L.D. Tripp, E.H. Schumacher, S.D. Keilholz, Short-time windows of correlation between large-scale functional brain networks predict vigilance intraindividually and interindividually. Hum. Brain Map. **34**(12), 3280–3298 (2013). https://onlinelibrary.wiley.com/doi/abs/10.1002/hbm.22140

74. G. Deco, J. Cruzat, J. Cabral, G.M. Knudsen, R.L. Carhart-Harris, P.C. Whybrow, N.K. Logothetis, M.L. Kringelbach, Whole-brain multimodal neuroimaging model using serotonin receptor maps explains non-linear functional effects of lsd. Current Biol. **28**(19), 3065–3074.e6 (2018). https://www.sciencedirect.com/science/article/pii/S0960982218310455

75. D. Wang, S. Liu, J. Warrell, H. Won, X. Shi, F.C.P. Navarro, D. Clarke, M. Gu, P. Emani, Y.T. Yang, M. Xu, M.J. Gandal, S. Lou, J. Zhang, J.J. Park, C. Yan, S.K. Rhie, K. Manakongtreecheep, H. Zhou, A. Nathan, M. Peters, E. Mattei, D. Fitzgerald, T. Brunetti, J. Moore, Y. Jiang, K. Girdhar, G.E. Hoffman, S. Kalayci, Z.H. Gümüş, G.E. Crawford, P. Consortium, P. Roussos, S. Akbarian, A.E. Jaffe, K.P. White, Z. Weng, N. Sestan, D.H. Geschwind, J.A. Knowles, M.B.

Gerstein, Comprehensive functional genomic resource and integrative model for the human brain. Science **362**(6420) (2018). https://science.sciencemag.org/content/362/6420/eaat8464

76. M.L. Kringelbach, G. Deco, Brain states and transitions: insights from computational neuroscience. Cell Rep. **32**(10), 108128 (2020). https://www.sciencedirect.com/science/article/pii/S2211124720311177

77. N. Tiwari, D.R. Edla, S. Dodia, A. Bablani, Brain computer interface: a comprehensive survey. Biol. Inspir. Cognit. Archit. **26**, 118–129 (2018). https://www.sciencedirect.com/science/article/pii/S2212683X18301142

78. J. DiGiovanna, L. Marchal, P. Rattanatamrong, M. Zhao, S. Darmanjian, B. Mahmoudi, J.C. Sanchez, J.C. Príncipe, L. Hermer-Vazquez, R. Figueiredo, J.A.B. Fortes, Towards real-time distributed signal modeling for brain-machine interfaces, in *Computational Science - ICCS 2007*, ed. by Y. Shi, G.D. van Albada, J. Dongarra, P.M.A. Sloot (Springer, Berlin, 2007), pp. 964–971

79. T. Mullen, C. Kothe, Y.M. Chi, A. Ojeda, T. Kerth, S. Makeig, G. Cauwenberghs, T.-P. Jung, Real-time modeling and 3d visualization of source dynamics and connectivity using wearable eeg, in *2013 35th Annual International Conference of the IEEE Engineering in Medicine and Biology Society (EMBC)* (2013), pp. 2184–2187

80. A. Nourmohammadi, M. Jafari, T.O. Zander, A survey on unmanned aerial vehicle remote control using brain-computer interface. IEEE Trans. Hum.-Mach. Syst. **48**(4), 337–348 (2018)

81. J. Wonderlick, D. Ziegler, P. Hosseini-Varnamkhasti, J. Locascio, A. Bakkour, A. van der Kouwe, C. Triantafyllou, S. Corkin, B. Dickerson, Reliability of mri-derived cortical and subcortical morphometric measures: effects of pulse sequence, voxel geometry, and parallel imaging. NeuroImage **44**(4), 1324–1333 (2009). https://www.sciencedirect.com/science/article/pii/S1053811908011518

82. C. Vollmar, J. O'Muircheartaigh, G.J. Barker, M.R. Symms, P. Thompson, V. Kumari, J.S. Duncan, M.P. Richardson, M.J. Koepp, Identical, but not the same: intra-site and inter-site reproducibility of fractional anisotropy measures on two 3.0t scanners. NeuroImage **51**(4), 1384–1394 (2010). https://www.sciencedirect.com/science/article/pii/S1053811910003332

83. S. Gu, F. Pasqualetti, M. Cieslak, Q.K. Telesford, A.B. Yu, A.E. Kahn, J.D. Medaglia, J.M. Vettel, M.B. Miller, S.T. Grafton, D.S. Bassett, Controllability of structural brain networks. Nat. Commun. **6**(1), 8414 (2015). https://doi.org/10.1038/ncomms9414

84. P. Ciuciu, J.-B. Poline, G. Marrelec, J. Idier, C. Pallier, H. Benali, Unsupervised robust nonparametric estimation of the hemodynamic response function for any fmri experiment. IEEE Trans. Med. Imaging **22**(10), 1235–1251 (2003)

85. D. Zoltowski, J. Pillow, S. Linderman, A general recurrent state space framework for modeling neural dynamics during decision-making, in *Proceedings of the 37th International Conference on Machine Learning*, Proceedings of Machine Learning Research, ed. by H.D. III, A. Singh, vol. 119. PMLR, 13–18 Jul 2020, pp. 11 680–11 691. http://proceedings.mlr.press/v119/zoltowski20a.html

86. J. Faskowitz, O. Sporns, Mapping the community structure of the rat cerebral cortex with weighted stochastic block modeling. Brain Struct. Funct. **225**(1), 71–84 (2020). https://doi.org/10.1007/s00429-019-01984-9

87. A.P. Alivisatos, M. Chun, G.M. Church, R.J. Greenspan, M.L. Roukes, R. Yuste, The brain activity map project and the challenge of functional connectomics. Neuron **74**(6), 970–974 (2012). https://www.sciencedirect.com/science/article/pii/S0896627312005181

88. B.O. Turner, E.J. Paul, M.B. Miller, A.K. Barbey, Small sample sizes reduce the replicability of task-based fmri studies. Commun. Biol. **1**(1), 1–10 (2018)

89. M. Mather, J.T. Cacioppo, N. Kanwisher, Introduction to the special section: 20 years of fmri-what has it done for understanding cognition? Perspect. Psychol. Sci. **8**(1), 41–43 (2013)

90. S. Kapur, A.G. Phillips, T.R. Insel, Why has it taken so long for biological psychiatry to develop clinical tests and what to do about it? Mol Psychiatry **17**(12), 1174–1179 (2012). https://doi.org/10.1038/mp.2012.105

91. H. Aerts, M. Schirner, T. Dhollander, B. Jeurissen, E. Achten, D. Van Roost, P. Ritter, D. Marinazzo, Modeling brain dynamics after tumor resection using the virtual brain. NeuroImage **213**, 116738 (2020). https://www.sciencedirect.com/science/article/pii/S1053811920302251

92. J. Zimmermann, A. Perry, M. Breakspear, M. Schirner, P. Sachdev, W. Wen, N. Kochan, M. Mapstone, P. Ritter, A. McIntosh, A. Solodkin, Differentiation of alzheimer's disease based on local and global parameters in personalized virtual brain models. NeuroImage: Clin. **19**, 240–251 (2018). https://www.sciencedirect.com/science/article/pii/S2213158218301268

93. K. Supekar, W. Cai, R. Krishnadas, L. Palaniyappan, V. Menon, Dysregulated brain dynamics in a triple-network saliency model of schizophrenia and its relation to psychosis. Biol. Psychiatry **85**(1), 60–69 (2019). Immune Mechanisms and Psychosis. https://www.sciencedirect.com/science/article/pii/S0006322318317153

94. R. Hultman, K. Ulrich, B.D. Sachs, C. Blount, D.E. Carlson, N. Ndubuizu, R.C. Bagot, E.M. Parise, M.-A.T. Vu, N.M. Gallagher, J. Wang, A.J. Silva, K. Deisseroth, S.D. Mague, M.G. Caron, E.J. Nestler, L. Carin, K. Dzirasa, Brain-wide electrical spatiotemporal dynamics encode depression vulnerability. Cell **173**(1), 166–180.e14 (2018). https://www.sciencedirect.com/science/article/pii/S0092867418301569

95. L.M. Hernandez, J.D. Rudie, S.A. Green, S. Bookheimer, M. Dapretto, Neural signatures of autism spectrum disorders: insights into brain network dynamics. Neuropsychopharmacology **40**(1), 171–189 (2015). https://doi.org/10.1038/npp.2014.172

96. E.T. Rolls, W. Cheng, J. Feng, Brain dynamics: the temporal variability of connectivity, and differences in schizophrenia and adhd. Transl. Psychiatry **11**(1), 70 (2021). https://doi.org/10.1038/s41398-021-01197-x

97. E. Müller, S. van Albada, J. Kim, P. Robinson, Unified neural field theory of brain dynamics underlying oscillations in parkinson's disease and generalized epilepsies. J. Theor. Biol. **428**, 132–146 (2017). https://www.sciencedirect.com/science/article/pii/S0022519317302886

98. J. Ou, L. Xie, C. Jin, X. Li, D. Zhu, R. Jiang, Y. Chen, J. Zhang, L. Li, T. Liu, Characterizing and differentiating brain state dynamics via hidden markov models. Brain Topogr. **28**(5), 666–679 (2015). https://doi.org/10.1007/s10548-014-0406-2

99. T. Kim, P. Hinds, Who should i blame? Effects of autonomy and transparency on attributions in human-robot interaction, in *ROMAN 2006 - The 15th IEEE International Symposium on Robot and Human Interactive Communication* (2006), pp. 80–85

100. M.R. Endsley, From here to autonomy: lessons learned from human–automation research. Hum. Fact. **59**(1), 5–27 (2017). pMID: 28146676. https://doi.org/10.1177/0018720816681350

101. T. Overbye, S. Saripalli, Path optimization for ground vehicles in off-road terrain (2021). arXiv:2101.00769

102. D. Chen, B. Zhou, V. Koltun, P. Krähenbühl, Learning by cheating, in *Proceedings of the Conference on Robot Learning*, Proceedings of Machine Learning Research, ed. by L.P. Kaelbling, D. Kragic, K. Sugiura, vol. 100. PMLR, 30 Oct–01 Nov (2020), pp. 66–75. http://proceedings.mlr.press/v100/chen20a.html

103. Y. Tang, C. Zhao, J. Wang, C. Zhang, Q. Sun, F. Qian, Perception and decision-making of autonomous systems in the era of learning: an overview (2020)

104. S.H. Vemprala, S. Saripalli, Collaborative localization for micro aerial vehicles. IEEE Access **9**, 63 043–63 058 (2021)

105. M. Lewis, H. Li, K. Sycara, Chapter 14 - Deep learning, transparency, and trust in human robot teamwork, in *Trust in Human-Robot Interaction*, ed. by C.S. Nam, J.B. Lyons (Academic, 2021), pp. 321–352. https://www.sciencedirect.com/science/article/pii/B9780128194720000149

106. M. Hu, T. Shealy, Application of functional near-infrared spectroscopy to measure engineering decision-making and design cognition: literature review and synthesis of methods. J. Comput. Civ. Eng. **33**(6), 04019034 (2019)

107. M. Ashtiani, M.A. Azgomi, A survey of quantum-like approaches to decision-making and cognition. Math. Soc. Sci. **75**, 49–80 (2015). https://www.sciencedirect.com/science/article/pii/S0165489615000165

108. M.R. Endsley, *Designing for Situation Awareness: An Approach to User-Centered Design* (CRC Press, 2016)

109. H. Berger, P. Gloor, *Hans Berger on the Electroencephalogram of Man: The Fourteen Original Reports on the Human Electroencephalogram.* EEG Journals Supplement (Elsevier Publishing Company, 1969). https://books.google.com/books?id=UAp7AQAACAAJ

110. A. Hodgkin, Edgar douglas adrian, baron adrian of cambridge. 30 november 1889-4 august 1977. Biograph. Mem. Fellows R. Soc. **25**, 1–73 (1979). http://www.jstor.org/stable/769841

111. H.H. Jasper, I.C. Nichols, Electrical signs of cortical function in epilepsy and allied disorders. Amer. J. Psych. **94**(4), 835–851 (1938). https://doi.org/10.1176/ajp.94.4.835

112. J.R. Knott, F.A. Gibbs, C.E. Henry, Fourier transforms of the electroencephalogram during sleep. J. Exp. Psychol. **31**(6), 465–477 (1942). https://doi.org/10.1037/h0058545

113. W.J. Freeman, W.g. walter: the living brain, in *Brain Theory*, ed. by G. Palm, A. Aertsen (Springer, Berlin, 1986), pp. 237–238

114. E.V. Evarts, Activity of neurons in visual cortex of the cat during sleep with low voltage fast eeg activity. J. Neurophysiol. **25**(6), 812–816 (1962). https://doi.org/10.1152/jn.1962.25.6.812

115. T. Enomoto, C. Ajmone-Marsan, Epileptic activation of single cortical neurons and their relationship with electroencephalographic discharges. Electroencephalogr. Clin. Neurophysiol. **11**(2), 199–218 (1959). https://www.sciencedirect.com/science/article/pii/0013469459900768

116. J.P. Rosenfeld, Applied psychophysiology and biofeedback of event-related potentials (brain waves): historical perspective, review, future directions. Biofeedback Self-regulation **15**(2), 99–119 (1990). https://doi.org/10.1007/BF00999142

117. G. McKhann, D. Drachman, M. Folstein, R. Katzman, D. Price, E.M. Stadlan, Clinical diagnosis of alzheimer's disease. Neurology **34**(7), 939 (1984). https://n.neurology.org/content/34/7/939

118. C. Davatzikos, F. Xu, Y. An, Y. Fan, S.M. Resnick, Longitudinal progression of Alzheimer's-like patterns of atrophy in normal older adults: the SPARE-AD index. Brain **132**(8), 2026–2035 (2009). https://doi.org/10.1093/brain/awp091

119. A.F. Goldszal, C. Davatzikos, D.L. Pham, M.X.H. Yan, R.N. Bryan, S.M. Resnick, An image-processing system for qualitative and quantitative volumetric analysis of brain images. J. Comput. Assist. Tomography **22**(5) (1998). https://journals.lww.com/jcat/Fulltext/1998/09000/An_Image_Processing_System_for_Qualitative_and.30.aspx

120. J. Dauwels, F. Vialatte, A. Cichocki, Diagnosis of alzheimers disease from eeg signals: where are we standing? Curr. Alzheimer Res. **7**(6), 487–505 (2010). http://www.eurekaselect.com/node/86614/article

121. R. Abiri, S. Borhani, E.W. Sellers, Y. Jiang, X. Zhao, A comprehensive review of EEG-based brain-computer interface paradigms. J. Neural Eng. **16**(1), 011001 (2019). https://doi.org/10.1088/1741-2552/aaf12e

122. E. Ruzich, M. Crespo-García, S.S. Dalal, J.F. Schneiderman, Characterizing hippocampal dynamics with meg: a systematic review and evidence-based guidelines. Hum. Brain Mapp. **40**(4), 1353–1375 (2019)

123. V.D. Calhoun, T.D. Wager, A. Krishnan, K.S. Rosch, K.E. Seymour, M.B. Nebel, S.H. Mostofsky, P. Nyalakanai, K. Kiehl, The impact of t1 versus epi spatial normalization templates for fmri data analyses. Hum. Brain Mapp. **38**(11), 5331–5342 (2017). https://onlinelibrary.wiley.com/doi/abs/10.1002/hbm.23737

124. B.U. Cowley, J. Korpela, Computational testing for automated preprocessing 2: practical demonstration of a system for scientific data-processing workflow management for high-volume eeg. Front. Neurosci. **12**, 236 (2018). https://www.frontiersin.org/article/10.3389/fnins.2018.00236

125. A. Delorme, T. Sejnowski, S. Makeig, Enhanced detection of artifacts in eeg data using higher-order statistics and independent component analysis. NeuroImage **34**(4), 1443–1449 (2007). https://www.sciencedirect.com/science/article/pii/S1053811906011098

126. A. Shoka, M. Dessouky, A. El-Sherbeny, A. El-Sayed, Literature review on eeg preprocessing, feature extraction, and classifications techniques. Menoufia J. Electron. Eng. Res **28**(1), 292–299 (2019)

127. F. Gargouri, F. Kallel, S. Delphine, A. Ben Hamida, S. Lehéricy, R. Valabregue, The influence of preprocessing steps on graph theory measures derived from resting state fmri. Front. Comput. Neurosci. **12**, 8 (2018). https://www.frontiersin.org/article/10.3389/fncom.2018.00008

128. N.R. Cook, Quantifying the added value of new biomarkers: how and how not. Diagn. Progn. Res. **2**(1), 14 (2018). https://doi.org/10.1186/s41512-018-0037-2

129. Z. Eaton-Rosen, F. Bragman, S. Bisdas, S. Ourselin, M.J. Cardoso, Towards safe deep learning: accurately quantifying biomarker uncertainty in neural network predictions, in *Medical Image Computing and Computer Assisted Intervention - MICCAI 2018*, ed. by A.F. Frangi, J.A. Schnabel, C. Davatzikos, C. Alberola-López, G. Fichtinger (Springer International Publishing, Cham, 2018), pp. 691–699

130. Y. Zhao, W. Zheng, D.Y. Zhuo, Y. Lu, X. Ma, H. Liu, Z. Zeng, G. Laird, Bayesian additive decision trees of biomarker by treatment interactions for predictive biomarker detection and subgroup identification. J. Biopharm. Stat. **28**(3), 534–549 (2018). pMID: 29020511. https://doi.org/10.1080/10543406.2017.1372770

131. M. Rashid, N. Sulaiman, A.P.P. Abdul Majeed, R.M. Musa, A.F.Ab. Nasir, B.S. Bari, S. Khatun, Current status, challenges, and possible solutions of eeg-based brain-computer interface: a comprehensive review. Front. Neurorobotics **14**, 25 (2020). https://www.frontiersin.org/article/10.3389/fnbot.2020.00025

132. M. de Dreu, I. Schouwenaars, G. Rutten, N. Ramsey, J. Jansma, Fatigue in brain tumor patients, towards a neuronal biomarker. NeuroImage: Clinical **28**, 102406 (2020). https://www.sciencedirect.com/science/article/pii/S2213158220302436

133. J.T. Lindgren, A. Merlini, A. Lécuyer, F.P. Andriulli, simbci-a framework for studying bci methods by simulated eeg. IEEE Trans. Neural Syst. Rehabil. Eng. **26**(11), 2096–2105 (2018)

134. J. Vorwerk, R. Oostenveld, M.C. Piastra, L. Magyari, C.H. Wolters, The fieldtrip-simbio pipeline for eeg forward solutions. BioMed. Eng. OnLine **17**(1), 37 (2018). https://doi.org/10.1186/s12938-018-0463-y

2.1 Electroencephalography as a Cognitive Biomarker

EEG is the most common high temporal resolution cognitive biomarker [1]. Current EEG electrodes can detect frequency content between 0.01 to around 100 Hz, with microvolt resolution. Because this work is especially focused on the *dynamics* of brain waves through biomarkers, the selected measure must have an appropriately high sampling rate for *dynamics*. Generally, sampling rapidly enough to distort the information in the output is uncommon, but sampling too slowly can easily impact the statistical variation in the output, which may hamper rigorous modeling approaches [2]. As a result, a preference is developed here for electrophysiological biomarkers, such as EEG or MEG, rather than metabolic biomarkers, such as fMRI. In the sections that follow, some of the important characteristics of EEG as they relate to dynamics are explored.

2.1.1 Relevant Characteristics of EEG for Dynamic Brain Wave Modeling

In Sect. 1.2.1.1, it was established that the link between electrical activity at the cellular level (i.e. brain waves) and EEG recordings is unclear at best. The first and most important difference between brain waves and EEG signals is that EEG signals are *referential*. Because EEG sensors are fundamentally electrodes, they rely on voltage differences between points. In the same way, a voltmeter depends on the placement of two leads, an EEG signal is the dynamic voltage difference between two electrodes. Unlike a voltmeter, however, there is no clear ground for brain wave activity. Note that this means a single channel EEG device requires two electrodes. The need for a relative reference has resulted in a great variety of methodological choices for researchers. Some EEG recordings are chained so that each electrode is relative to its neighboring electrode laterally or longitudinally. Figure 2.1 shows

© The Author(s), under exclusive license to Springer Nature Switzerland AG 2023

T. D. Griffith et al., *A Modal Approach to the Space-Time Dynamics of Cognitive Biomarkers*, Synthesis Lectures on Biomedical Engineering, https://doi.org/10.1007/978-3-031-23529-0_2

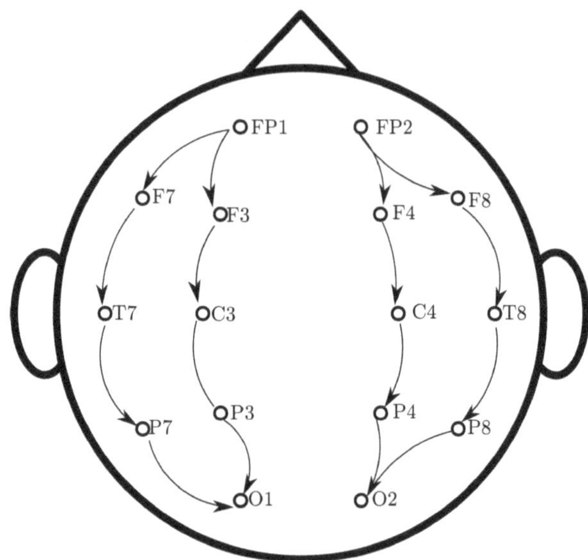

Fig. 2.1 An example of longitudinal referencing with electrode locations and names according to the International 10–20 system standard

a simple example of longitudinal bipolar referencing. In this example, each arrow represents a dynamic EEG signal, which is the voltage difference between two given electrodes. Still, other recordings are referenced to an electrode at the top center of the skull, while others are referenced to the mathematical average or to an electrode on the ear. An overview of the different referencing methods and a discussion of how the referencing method impacts modeling outcomes can be found in [3]. Here again, a fundamental issue with the study of brain waves is revealed. Because of the uncertainty associated with brain waves, it is not clear that one reference frame is better than another, which can lead to confusing and often conflicting results [4]. From an information theoretic standpoint, the entropy of an EEG recording should not change based on the referencing system, however [5].

Further, while brain waves flow in a deterministic fashion along electrochemical pathways in the brain, the mass neuronal activity passes through several biological filters before it is detected by EEG sensors. Electrical activity in the cortex must pass through the pia mater, dura mater, spinal fluid, skull, and skin before it can be detected by EEG electrodes. Each of these filters introduces a smearing effect on the electrical information emitting from the cortex [6]. This, in turn, may induce constructive and destructive interference in the brain wave information. As a result, each electrode detects the joint activity of about a billion cortical neurons, and EEG measures are said to have poor spatial resolution [6]. Subcortical structures have no influence on EEG measures of brain waves, even though those structures also generate and transmit brain waves [7]. Crucially, this smearing, which is often termed volume conduction, distorts the temporal information in the EEG signal relative to the underlying source brain wave [8]. This decrease in temporal EEG resolution as a result of volume conduction is well studied, but often ignored in the analysis of EEG brain waves. Overall, in the study of brain wave dynamics with EEG measures, it is important to remember that the channels are relative to some reference and that the resultant signal has passed through several biological filters by volume conduction. EEG signals do not measure brain waves directly.

2.1.2   Statistical Properties of EEG Measures

Naturally, these characteristics influence the statistical properties of EEG measures. The most obvious of these is the increased covariance between neighboring EEG electrodes as a result of volume conduction. Figure 2.2 shows a comparison of four channels of EEG in a

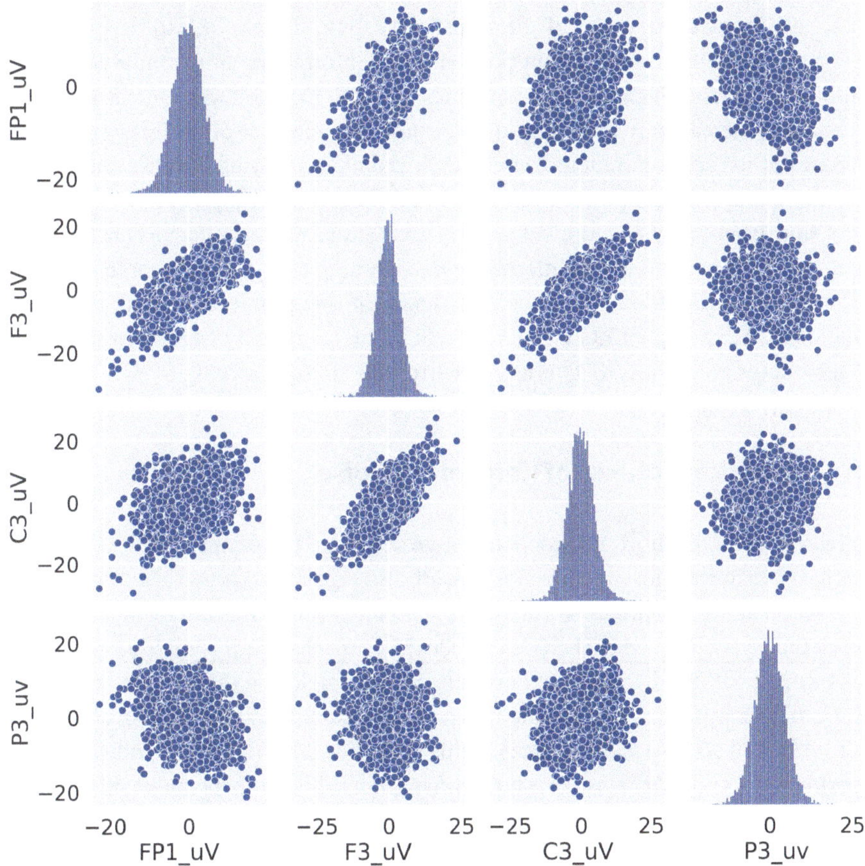

Fig. 2.2 Pair plot of four detrended EEG channels (FP1, F3, C3, P3) in the 10–20 standard. The univariate distribution of each channel is shown on the diagonal, while the off-diagonal plots show the correlation of two different channels in a bivariate distribution. Channels with the exact same information would fall on a straight line with a slope of one

pair plot. Notice that the channels which are spatially close to one another tend to fall more neatly along a line. For example, C3 and F3 are spatially close, so the bivariate distribution falls more neatly along the diagonal, while F3 and P3 are spatially distant, so their bivariate distribution is less correlated. That is because volume conduction "smears" the mass neuronal activity of the brain, EEG sensors near one another spatially will share much of the same information. While this "smearing" is not particularly surprising, it demonstrates the need for brain wave models of EEG activity to incorporate spatial dynamics. *EEG channels are not wholly independent sources of information.*

Second, EEG recordings of brain waves have nonstationary statistics. That is, the statistical moments, such as the mean, variance, and skew are functions of time and space. Because stationary statistics ground so many of our analytical methods, researchers historically treated the "non-stationarity" of EEG signals as a source of noise, discarding it in favor of just the stationary part of the signal with some success [9, 10]. Recognize, however, that by discarding a significant portion of the information in the measure, there may be difficulties interpreting modeling outcomes [11, 12]. Compounding the difficulty of this problem, it is increasingly clear that the nonstationary portion of an EEG signal and the stationary portion are *both* fundamental to the dynamic activity of brain waves [13]. Large-scale neuronal activity arises from the synchronized sudden firings of billions of cells, so it is not surprising that the macro behavior is nonstationary. This effect has been especially demonstrated when brain wave patterns are transitioning between metastable patterns [14]. Here, this work seeks to embrace the complexity of this problem, treating the nonlinear, nonstationary behavior of EEG signals. To add to the complexity of brain wave dynamics, EEG recordings are subject to a variety of corrupting noise sources.

2.1.3   Biological Sources of Corrupting Noise

Brain waves are a result of electrochemical processes in the human body. Broadly, many anatomical structures, most notably muscles, are also regulated through electrochemical processes. These biological artifacts from other anatomical structures also present as electrical waves, which are observed by EEG electrodes because the scalp is highly conductive and EEG electrodes do not discriminate between electrical fields on the basis of their source. If these biological artifacts are not treated, they can obscure the modeling process. For example, the electrical activity of the heart is highly periodic and detectable with EEG. However, this periodic electrical signal is not a brain wave, and it should not be modeled as such. This is especially important when the artifacts have frequency content that may overlap with the frequency content of brain waves. This example leads us into a discussion of the most critical biological artifacts for EEG observations: muscular, cardiac, and ocular.

2.1.3.1 Muscular Artifacts

A discerning reader might note that each of these biological sources of noise is the result of muscular activity. The heart is a muscle, and muscles are responsible for both eye blinks and eye movement. The distinction, in this case, is in how the specific anatomical structure gives rise to a specific artifact. Muscular artifacts include myogenic activity from a variety of muscle groups which are characterized by wide spectrum disturbances, perturbing all the standard EEG bands [15]. Jaw, neck, and limb movement are especially influential on EEG observations, and are especially corrupting of 15–30 Hz brain waves and 0.1–2 Hz brain waves [16, 17]. As a result, many EEG studies historically limit subjects to a stationary position, but more recent results suggest that independent component analysis through blind source separation is well suited to removing broad-spectrum muscular artifacts from EEG recordings [18]. In a real-time application, it is often necessary and effective to simply suppress the EEG recording in the presence of muscular artifacts [19]. Muscular artifacts from myogenic activity corrupt EEG recordings through volume conduction across a wide spectral range.

It is worth recognizing that muscular activity is a valid biomarker in its own right. Electromyography (EMG) seeks to observe myogenic activity and is widely used to diagnose nerve or muscle disorders [20]. While EMG recordings are not brain waves, they may be related to brain waves; a secondary observable of the system as a whole [21]. A modeling method considering multiple data streams may reveal new insights into the influence of brain waves on motor control.

2.1.3.2 Cardiac Artifacts

Cardiac artifacts are present in EEG recordings as a result of the electrical activity in the heart. Fortunately, the artifact generated by a beating heart tends to be low amplitude and very repetitive [16]. Because of its periodicity, it may be easily confused for a brain wave. However, a reference waveform for this repetitive artifact is usually available, and robust processing algorithms are well established for the removal of cardiac artifacts [22].

Practically, if an EEG electrode is placed too near a blood vessel, the pumping action of the blood generates local periodic waves that again may easily be confused with brain waves [23]. Fortunately, a direct link between the cardiac waveform discussed in the previous paragraph and these pulse artifacts has been established [24]. Accordingly, these signals too may be easily treated, but it is generally recommended to place electrodes away from obvious vessels when possible [25].

2.1.3.3 Ocular Artifacts

Ocular artifacts arise from both eye movement and eye blinks. Eye movement involves muscular activity that is spatially local at the frontal electrodes and is of moderate amplitude. Eye blinking, on the other hand, tends to be abrupt and of greater amplitude. From an energy perspective, the electrical power needed to move an eyelid is much greater than that required for the steady oscillation of brain waves, so eye blinks tend to overwhelm and wash out EEG signals. As a result, having reference waveforms for eye movement for cancelation is advisable [26], but there is bi-directional contamination between brain waves and ocular artifacts [27]. This bi-directional contamination necessitates careful filtering of eye movement artifacts, which is well studied [28, 29].

2.1.4 Inorganic Sources of Corrupting Noise

Finally, EEG electrodes are sensitive enough to detect environmental sources of corrupting noise, which are inorganic. One can argue that the biological artifacts are part of the whole brain system, but these inorganic sources of noise must absolutely be canceled. Fortunately, inorganic artifacts are the easiest sources of EEG noise to correct. The primary source of inorganic noise in EEG observations is the background hum of power leads and wall outlets, which oscillate at either 50 Hz or 60 Hz, depending on a given country's power grid. EEG recordings in the United States, for example, show a clear spectral artifact at 60 Hz. This noise can be easily dealt with by implementing a simple filter at the appropriate frequency. Additionally, as with all electromagnetic measuring devices, EEG observations can be corrupted by magnetic interference, so electrode leads should be well shielded and as short as possible. A detailed review of these inorganic sources of noise is presented in [30], where they are shown to be manageable.

Overall, while these sources of corrupting noise are not overwhelming, they must be treated. A comprehensive review of artifacts and their treatments is presented in [15]. With regard to biological artifacts, this work strongly argues for minimal manipulation of the raw data. Ideally, one may perfectly insulate EEG recordings from any outside influence, but at present, this is not possible [25]. Retaining as much data as possible ensures the least amount of relevant information in the signal is destroyed [16]. Ultimately, EEG data is known to contain periodic trends and noise. Models should be rooted in the original analysis of brain waves; connected to the analysis of periodic brain wave patterns. Highly derived preprocessing methods too easily become uninterpretable and untestable, not to mention empirically meaningless [31]. The temptation of fitting data to a hypothesis must be avoided.

Section 2.1 has outlined a variety of technical considerations for the modeling of brain wave dynamics via EEG recordings. Throughout it has been established that:

- EEG signals are relative to a reference, which may take many different forms;
- EEG signals have nonstationary statistics;
- EEG signal covariance is inversely proportional to the spatial distance between two sensors;
- EEG electrodes detect activity from the eyes, heart, and muscles that may look like brain waves; and
- EEG electrodes detect activity from inorganic sources, such as electrical outlets.

As a result of all these factors, much debate has been generated analyzing whether EEG signals are highly nonlinear and deterministic, or completely stochastic. Regardless of the underlying dynamics, EEG signals are "noisy". *While long time series recordings of EEG data are clearly nonstationary, it has been shown that they are time-invariant over shorter periods of time on the order of minutes* [15]. Some local linearity may be determined for many nonlinear dynamical systems in engineering, but it is encouraging to see this result reproduced for a biological system (i.e. brain wave dynamics). Accordingly, the tools of dynamics for such engineering systems become relevant to brain wave dynamics, as was introduced briefly in Sect. 1.4. With the requisite background knowledge in hand, this use of engineering dynamics principles for brain wave analysis is further expanded and motivated.

2.2   A Canonical Approach to the Analysis of Brain Wave Dynamics

Analysis of joint spatio-temporal systems has been widely considered in engineering dynamics [32]. Spatio-temporal brain wave dynamics contain structured linear and nonlinear behavior, as well as unstructured random noise. The approach in this work is especially focused on elegantly treating each of these dynamic effects. As we will show in Sect. 6.3, the proposed approach operates exactly in this way, filtering the random noise, while estimating the linear and nonlinear dynamics. First, we treat linear dynamics. Ordinary differential equations (ODEs) can canonically represent a given linear system's dynamics in state space as

$$\begin{cases} \dot{x}(t) = Ax(t) + Bu(t) \\ y(t) = Cx(t). \end{cases} \tag{2.1}$$

Internal plant states evolve in the n-dimensional vector space $X \equiv \mathbb{C}^n$. The input is in the m-dimensional space $U \equiv \mathbb{C}^m$. The output is in the p-dimensional space $Y \equiv \mathbb{C}^p$. y contains the measurements of a prescribed set of sensors as a time-variant vector. In this work, y is a vector with each channel of EEG data in a different row. The number of rows is equal to the number of recorded channels. y is the most accessible representation of the information in a dynamical system. However, we must acknowledge that our sensors may not capture all the dynamics in the system. This is especially true of EEG measures which have limited spatial resolution and are smeared by cerebrospinal fluid (CSF) [33]. We model these *internal* dynamics with the state vector $x \in X$. x is frequently inaccessible directly and its dimension can be difficult to determine, so we often extract x with some optimal estimation technique [34]. In the context of brain waves, we call x a *cognitive state*, since it is the internal state of the system which gives rise to the observable brain waves y. $u \in U$ is the input to the system. In a mechanical system, the inputs can often be precisely controlled for improved analysis. Here we must accept that a mathematical representation of the inputs to the brain wave system remains elusive and unknown. The unknown inputs act through the matrix B, which distributes each unknown input across the appropriate internal states.

A is a matrix describing the coupled ordinary differential equation for the internal system dynamics x. It models the time change behavior of x, and frequently contains the relevant constitutive properties of the system. Finally, C represents the linear transformation from internal states x to the measurable output space y. It encodes information related to a given sensor setup for a system. Note crucially, that if the pair (A, C) is not observable, we will not be able to reconstruct x entirely from y alone.

Others have used ODEs to model the dynamics of biomarkers. In [35], a semiparametric model was applied to patients affected by alcohol dependence. [36] used first order ODEs to estimate the emergence of cognitive decline from biomarkers. Finally, a one-compartment model was used in [37] for the detection of tumors. In the case of EEG dynamics, A, B, C, x and u are all unknown. Here, we are interested in recovering the physical structure of the state space model without restricting ourselves to a specific parametric ODE. We hypothesize that appropriate estimates of A, B, C, x, and u are relevant to cognitive modeling outcomes.

Our modeling approach is segmented into two parts. Formulating the triple (A, B, C) from output measurements y alone remains an unsolved problem in system identification, especially for nonlinear systems, so we must reduce the complexity of the problem. Consider the true state space model that describes the linear dynamics of an observed set of brain wave measurements

$$\text{TRUE BRAIN WAVE PLANT:} \begin{cases} \dot{x} = Ax + Bu + v_x \\ y = Cx. \end{cases} \tag{2.2}$$

v_x is a bounded but unknown disturbance in the plant state. Additionally, the input also has a bounded but unknown disturbance v_u, which enters the system through the plant's input. Here the signals (v_x, v_u) represent uncertainties in our knowledge of the system model and of the input model we are using to approximate the unknown inputs. Whatever these uncertain signals represent: parametric knowledge, nonlinearities, or noise corruptions, we want to be able to assess their influence and build into our adaptive state estimation some ability to reduce their cumulative effect. It is reasonable to expect some finite bound on these uncertainties. We express this bound as

$$v = \begin{bmatrix} v_x \\ v_u \end{bmatrix} \tag{2.3}$$

$$||v|| \leq M_v < \infty. \tag{2.4}$$

There are a variety of system identification techniques, explored further in Sect. 3.1.3, to estimate the matrix pair (A, C) in the presence of an unknown input. These techniques mitigate the effect of the unknown input and synthesize a description of the system dynamics at regular operating conditions. We introduce the distinction between A and A_m, because the model of the internal dynamics A_m will be especially susceptible to parametric uncertainty and some of the information in the unknown input could perturb A_m. We set aside the exact methods for synthesizing this matrix pair until Sect. 3.1.3 and continue assuming we have a reasonable approximation of (A, C) which we denote (A_m, C). We assume the estimate of C is reasonably good, since (A, C) cannot be uniquely determined with the output alone as the two are coupled [38]. Compensating for parametric uncertainty in either matrix treats the performance of the whole model, and it is standard practice to consider perturbation of A [39]. Our model of the brain wave system at this stage takes the form

$$\text{UNCERTAIN LINEAR MODEL:} \begin{cases} \dot{x}_m = A_m x \\ y_m = C x_m. \end{cases} \tag{2.5}$$

In the context of this work, (A_m, C) yields a valid description of the *linear* brain wave dynamics in an observed EEG recording y, uncoupled from the effects of external inputs and ignoring nonlinear effects. While the effect of the input is important, the constitutive properties of the *linear system* are completely contained in (A, C). Brain waves are known to contain linear behavior, nonlinear behavior, and random noise, so getting an initial estimate of the linear behavior is computationally feasible and important. The effective synthesization of (A, C) from y is the first step in our canonical modeling approach. We recognize that there are other significant dynamics at play, and at this stage, we should not expect particularly high-fidelity models. However, simplified models of this form still have useful diagnostic information, as we will show in Sect. 4.1. In the second step of our approach, we treat the nonlinear plant effects and unknown input with an adaptive estimator.

2.2.1   Treating the Nonlinear Effects of Brain Waves

As introduced in Sect. 1.4, because brain waves exhibit nonlinear, nonstationary dynamics, a single model of the form Eq. (2.1) is unlikely to describe the global brain wave dynamics. Much of modern control theory is dedicated to analyzing and stabilizing nonlinear systems. Often, engineers linearize the plant dynamics around a discrete set of known operating conditions. Each linearization takes the form of Eq. (2.1), and then a variety of algorithms may switch between linearizations as needed to improve the model accuracy or performance outcomes [40–42]. Of course, it is difficult to assess which operating points are important to brain waves. Therefore, we introduce a novel adaptive estimation scheme in Chap. 5 to simultaneously treat nonlinear plant dynamics and estimate the unknown system input. This adaptive estimator updates the physical structure of (A, C) to better match the observed plant dynamics. The adaptive estimator is highly nonlinear and modifies (A, C) to account for parametric uncertainty and nonlinear effects. Simultaneously, the estimator forms a representation of the system input u which we call $\hat{u}$. We will hypothesize and then validate that the unknown system input acts evenly on the spatial brain wave dynamics, which implies that B is a matrix of ones. Here, we present this adaptive approach to real time brain wave estimation as

$$\text{ADAPTIVE BRAIN WAVE ESTIMATOR:} \begin{cases} \dot{\hat{x}} = \big(A_m + BL(t)C\big)\hat{x} + B\hat{u} + K_x e_y; \\ \hat{y} = C\hat{x}. \end{cases} \quad (2.6)$$

Again, and crucially $A \neq A_m$. A, B, C, and K_x are all static matrices of appropriate size. We may only use the observable information in the system (the output y) to correct A_m. Through adaptive output feedback, $A_m + BL(t)C$ approximates A, even when faced with nonlinear dynamics. In this view, the term $BL(t)C$ updates A_m to track the nonlinear brain wave dynamics. Each $L(t)$ yields a different linear model of the form Eq. (2.1). We will demonstrate that Eq. (2.6) generates estimates $\hat{y}$ of the measured brain wave signal y that are useful to cognitive modeling outcomes.

2.3   Modal Analysis of State Space Brain Wave Models

The methods for developing high-fidelity models of processes described by Eq. (2.1) are well developed which encourages us to apply them to brain wave data. Unmodified, however, it may be difficult to extract useful modeling features from Eq. (2.6) even if the dynamic brain wave model is high fidelity. The dimension of A_m can be quite large and does not yield a physically interpretable tool for the analysis of brain wave dynamics. Certainly, models of this form readily supply information related to standard dynamic system theory, such as stability, observability, and controllability, but this information may not be easily converted to cognitive modeling outcomes. Therefore, as an additional step in our modeling procedure, we perform a modal transformation of these state-space models which yields a discrete set of *spatial brain wave patterns. Each of these spatial patterns has an associated oscillating frequency, which aligns with the known behavior of brain waves.* These patterns are especially useful for cognitive modeling and developing physical insight into brain waves from our modeling approach. That is, while models of the form Eq. (2.1) describe the measured brain wave well, a modal view of the dynamics better accounts for the body of existing knowledge on brain waves even though the systems are mathematically equivalent.

2.3.1   Modes Jointly Capture Space Time Dynamics

Recall from Sect. 1.3 that brain waves are known to have spatio-temporal patterns. Systems that can be described by Eq. (2.1) admit eigenmodes. Even though our estimator takes the form Eq. (2.6), it is estimating the true brain wave plant represented by Eq. (2.1) Eigenmodes represent the data in an analytical basis set that organizes the observed behavior of the system into distinct oscillating frequencies and their associated mode shapes, which we have hypothesized are relevant to the dynamics of brain waves. A modal decomposition is necessarily a linear transform, which models the time behavior of modal amplitudes $\eta(t)$ instead of the internal states $x(t)$

$$x = V\eta. \tag{2.7}$$

Each column of V is an eigenvector of A

$$V = \begin{bmatrix} v_1 & v_2 & \ldots & v_n \end{bmatrix}. \tag{2.8}$$

If the n eigenvectors are linearly independent, the modal transformation matrix V has an inverse. Then, V diagonalizes A into a matrix of system eigenvalues λ_i

$$VAV^{-1} = \Lambda = \begin{bmatrix} \lambda_1 & 0 & \ldots & 0 \\ 0 & \lambda_2 & \ldots & 0 \\ \vdots & \vdots & \ddots & \vdots \\ 0 & 0 & \ldots & \lambda_n \end{bmatrix}. \tag{2.9}$$

Accordingly, any system state can be written as a weighted linear combination of modes weighted by α_i

$$x(t) = \sum_{i=1}^{n} \alpha_i v_i e^{\lambda_i t} \tag{2.10}$$

$$y(t) = \sum_{i=1}^{n} \alpha_i C v_i e^{\lambda_i t}. \tag{2.11}$$

The dynamics of the system have now been sorted into a set of spatio-temporal eigenmodes. Each v_i is a mode shape, with an associated eigenvalue λ_i. When considered together, v_i and λ_i form an *eigenmode*. The modal decomposition results in a modified state-space model of the form

$$\begin{cases} \dot{\eta}(t) = \Lambda\eta(t) + V^{-1}Bu(t) \\ y(t) = CV\eta(t). \end{cases} \tag{2.12}$$

η is simply the modal amplitude of each eigenmode at a given time. In this form, the states η are uncoupled in Λ but are coupled through the input and the output. The observed EEG measure is a weighted linear combination of modes. Notice that while we make the distinction between LTI models of the form (A, B, C) and $(\Lambda, V^{-1}B, CV)$, they are mathematically equivalent.

2.3.2   Analytical Relevance of Eigenmodes

With an overview of the mathematics for modal analysis of neural activity complete, we turn to some properties of modes that are relevant to cognitive modeling outcomes. We discuss the mechanical view of each property because it informs the reason for including the property and aids the interpretation of EEG modes. Specifically, modes are characterized by frequency, damping, mode shape, and complexity. Here, these properties are mathematical constructs that aid the description of the observed EEG dynamics. *Frequency* is a relatively simple property. EEG data has rich spectral content, and research has linked frequency specific features with cognitive outcomes [43, 44] Each mode has a distinct oscillating frequency, which directly corresponds to the spectral content in the EEG data since a weighted sum of modes recreates the original data.

The *damping coefficient* of a mode is a convenient measure of how much energy is dissipated from a system. The damping coefficient is frequently expressed as a percentage. A mode with a damping coefficient of 100% is critically damped. A mode with 0% damping oscillates indefinitely. In the view of EEG modes, modes with higher damping are transient, and their effects are limited in the temporal domain. This suggests a physiological change in the brain's spectral content, which is often attributed to brain network evolution [45]. Curiously, our models allow for the identification of modes with negative damping coefficients. Negative damping is an undesirable property in mechanical structures or circuits because the mode amplitude becomes unbounded over time. However, we accept negatively damped modes in EEG dynamics, which should be interpreted as the brain adding energy in that frequency band, because EEG measures are not known to diverge.

The *mode shape* has the richest modal information. For each measurement channel, there is an associated scalar that indicates the maximum amplitude of that channel for each mode. We may choose to think of the mode shape as a vector describing the relative contribution of each channel to the overall observed dynamics y for a given frequency.

2.3.2.1 Complexity as a Measure of Spatial Dependence

While the mode shape indicates the relative contribution of each spatial channel to the overall dynamics, it does not give an indication of whether the relative contributions are oscillating together. That is, while they may have the same temporal frequency, two spatial elements may not reach their maxima and minima together. This behavior, which gives rise to *traveling waves*, indicates that the spatial locations are out of the temporal phase. Again, in the view of mechanical structures, a beam may have a non-proportional damping element. This results in some elements of the beam lagging the rest of the beam. *Complexity* measures how much of the beam is lagging the initial excitation. Mathematically, modal complexity is defined as

$$C_r = 1 - \frac{(S_{xx} - S_{yy})^2 + 4S_{xy}^2}{(S_{xx} + S_{yy})^2}, \tag{2.13}$$

where

$$S_{xx} = \text{Re}(v_i^T)\,\text{Re}(v_i), \tag{2.14}$$

$$S_{yy} = \text{Im}(v_i^T)\,\text{Im}(v_i), \tag{2.15}$$

$$S_{xy} = \text{Re}(v_i^T)\,\text{Im}(v_i). \tag{2.16}$$

A mode that is completely in phase, reaching its maximum and minimum amplitude at every spatial location together, has a complexity value of 0%. Modes for physical systems tend to be low complexity because they are constrained mechanically and remain in phase. In the view of EEG modes, we allow for significant complexity, because the oscillatory behavior in the electrical activity of one part of the brain does not necessarily correspond to subsequent activity in another. For example, a sudden burst of activity in the frontal lobe is unlikely to be matched by activity in the cerebellum. The subsequent mode would necessarily be complex. Complexity in the brain's modes should be interpreted as a measure of the correlation between different sections of the cerebrum.

Accordingly, the whole literature on optimal state estimation becomes available to us, and we can track the brain wave dynamics in real time, while keeping the error between the estimated output limited to some radius centered at zero. The "features" v_i are generated in a canonical fashion and are directly tied to the measured data y through least squares projections and linear transformations only. Any result from the state-space model can be directly tied back to the data.

2.4   System Identification Tools for Brain Wave Analysis

This chapter has established an overview of the modeling considerations for brain wave dynamics and how each of these considerations forms our modeling approach. Overall, brain wave recordings, which are represented here in the vector y, have structured information and random noise. The data is relative to a reference which may vary from laboratory to laboratory and is sensitive to a variety of biological and artificial noise sources. *As a result, our modeling approach is to individually model each of these relevant effects. We filter the noise, model the linear behavior, and compensate for the nonlinear behavior with a nonlinear estimator. In order to gain physical insight into brain waves for cognitive modeling, these models are transformed from the standard state-space formulation to modal canonical form, which represents the system as a weighted linear combination of spatial patterns which form a basis for the observed dynamics.* We have thus far obscured the theoretical considerations to generate both Eq. (2.5) and Eq. (2.6). With this overview complete, we wish to focus on the synthesization and analysis of models in the form of Eq. (2.5) in Chap. 3.

References

1. D. Mantini, L. Marzetti, M. Corbetta, G.L. Romani, C. Del Gratta, Multimodal integration of fmri and eeg data for high spatial and temporal resolution analysis of brain networks. Brain Topogr. **23**(2), 150–158 (2010). https://doi.org/10.1007/s10548-009-0132-3
2. A. Horch, A.J. Isaksson, Assessment of the sampling rate in control systems. Control Eng. Prac. **9**(5), 533–544 (2001). https://www.sciencedirect.com/science/article/pii/S0967066101000156
3. L.R. Trambaiolli, A.C. Lorena, F.J. Fraga, P.A.M.K. Kanda, R. Nitrini, R. Anghinah, Does eeg montage influence alzheimer's disease electroclinic diagnosis? Int. J. Alzheimer's Dis. **2011**, 761891 (2011). https://doi.org/10.4061/2011/761891
4. D.A. Kaiser, Qeeg: state of the art, or state of confusion. J. Neurother. **4**(2), 57–75 (2000). https://doi.org/10.1300/J184v04n02_07
5. C.E. Shannon, A mathematical theory of communication. Bell Syst. Techn. J. **27**(3), 379–423 (1948)
6. S. van den Broek, F. Reinders, M. Donderwinkel, M. Peters, Volume conduction effects in eeg and meg. Electroencephalogr. Clin. Neurophysiol. **106**(6), 522–534 (1998). https://www.sciencedirect.com/science/article/pii/S0013469497001478
7. P. Federico, J.S. Archer, D.F. Abbott, G.D. Jackson, Cortical/subcortical bold changes associated with epileptic discharges. Neurology **64**(7), 1125–1130 (2005). https://n.neurology.org/content/64/7/1125
8. B. Burle, L. Spieser, C. Roger, L. Casini, T. Hasbroucq, F. Vidal, Spatial and temporal resolutions of eeg: is it really black and white? a scalp current density view. Int. J. Psychophysiol. **97**(3), 210–220 (2015). On the benefits of using surface Laplacian (current source density) methodology in electrophysiology. https://www.sciencedirect.com/science/article/pii/S0167876015001865
9. W. Klimesch, *Event-related Band Power Changes and Memory Performance* (Elsevier, 1999), pp. 161–178
10. E. Başar, M. özgören, S. Karakaş, C. Başar-eroglu, Super-synergy in the brain: the grandmother percept is manifested by multiple oscillations. Int. J. Bifur. Chaos **14**(02), 453–491 (2004). https://doi.org/10.1142/S0218127404009272
11. A.A. Fingelkurts, A.A. Fingelkurts, C.M. Krause, M. Sams, Probability interrelations between pre-/post-stimulus intervals and erd/ers during a memory task. Clin. Neurophysiol. **113**(6), 826–843 (2002). https://www.sciencedirect.com/science/article/pii/S1388245702000585
12. A. Effern, K. Lehnertz, G. Fernández, T. Grunwald, P. David, C. Elger, Single trial analysis of event related potentials: non-linear de-noising with wavelets. Clin. Neurophysiol. **111**(12), 2255–2263 (2000). https://www.sciencedirect.com/science/article/pii/S1388245700004636
13. A.I.a. Kaplan, The nonstability of the EEG: a methodological and experimental analysis. Usp Fiziol Nauk **29**(3), 35–55 (1998)
14. A.Y. Kaplan, A.A. Fingelkurts, A.A. Fingelkurts, S.V. Borisov, B.S. Darkhovsky, Nonstationary nature of the brain activity as revealed by eeg/meg: methodological, practical and conceptual challenges. Signal Proc. **85**(11), 2190–2212 (2005). Neuronal Coordination in the Brain: A Signal Processing Perspective. https://www.sciencedirect.com/science/article/pii/S0165168405002094
15. J.A. Urigüen, B. Garcia-Zapirain, EEG artifact removal—state-of-the-art and guidelines. **12**(3), 031001 (2015). https://doi.org/10.1088/1741-2560/12/3/031001
16. L. Sörnmo, P. Laguna, *Bioelectrical Signal Processing in Cardiac and Neurological Applications*, vol. 8. (Academic, 2005)

17. I. Goncharova, D. McFarland, T. Vaughan, J. Wolpaw, Emg contamination of eeg: spectral and topographical characteristics. Clin. Neurophysiol. **114**(9), 1580–1593 (2003). https://www.sciencedirect.com/science/article/pii/S1388245703000932

18. R. Vigario, E. Oja, Bss and ica in neuroinformatics: from current practices to open challenges. IEEE Rev. Biomed. Eng. **1**, 50–61 (2008)

19. J. Ma, P. Tao, S. Bayram, V. Svetnik, Muscle artifacts in multichannel eeg: characteristics and reduction. Clin. Neurophysiol. **123**(8), 1676–1686 (2012). https://www.sciencedirect.com/science/article/pii/S1388245711009084

20. M.B.I. Reaz, M.S. Hussain, F. Mohd-Yasin, Techniques of emg signal analysis: detection, processing, classification and applications. Biol. Proc. Online **8**(1), 11–35 (2006). https://doi.org/10.1251/bpo115

21. D.N.F.P. Damit, S.M.N. Arosha Senanayake, O.A. Malik, N.J. Tuah, Neuromuscular fatigue analysis of soldiers using dwt based emg and eeg data fusion during load carriage, in *Intelligent Information and Database Systems*. ed. by N.T. Nguyen, S. Tojo, L.M. Nguyen, B. Trawiński (Springer International Publishing, Cham, 2017), pp. 602–612

22. M.B. Hamaneh, N. Chitravas, K. Kaiboriboon, S.D. Lhatoo, K.A. Loparo, Automated removal of ekg artifact from eeg data using independent component analysis and continuous wavelet transformation. IEEE Trans. Biomed. Eng. **61**(6), 1634–1641 (2014)

23. B.J. Fisch, R. Spehlmann, *EEG Primer: Basic Principles of Digital and Analog EEG* (Elsevier Health Sciences, 1999), pp. 23–27

24. S.R. Benbadis, D. Rielo, L. Huszar, F. Talavera, N. Alvarez, P. Barkhaus, Eeg artifacts. Distribution **12**, 1–23 (2010)

25. P. Anderer, S. Roberts, A. Schlögl, G. Gruber, G. Klösch, W. Herrmann, P. Rappelsberger, O. Filz, M.J. Barbanoj, G. Dorffner, B. Saletu, Artifact processing in computerized analysis of sleep eeg – a review. Neuropsychobiology **40**(3), 150–157 (1999). https://www.karger.com/DOI/10.1159/000026613

26. T.T. Pham, R.J. Croft, P.J. Cadusch, R.J. Barry, A test of four eog correction methods using an improved validation technique. Int. J. Psychophysiol. **79**(2), 203–210 (2011). https://www.sciencedirect.com/science/article/pii/S0167876010007221

27. S. Romero, M.A. Mañanas, M.J. Barbanoj, A comparative study of automatic techniques for ocular artifact reduction in spontaneous eeg signals based on clinical target variables: a simulation case. Comput. Biol. Med. **38**(3), 348–360 (2008). https://www.sciencedirect.com/science/article/pii/S0010482507001850

28. G.L. Wallstrom, R.E. Kass, A. Miller, J.F. Cohn, N.A. Fox, Automatic correction of ocular artifacts in the eeg: a comparison of regression-based and component-based methods. Int. J. Psychophysiol. **53**(2), 105–119 (2004). https://www.sciencedirect.com/science/article/pii/S0167876004000510

29. G. Gratton, Dealing with artifacts: the eog contamination of the event-related brain potential. Behav. Res. Methods Instrum. & Comput. **30**(1), 44–53 (1998). https://doi.org/10.3758/BF03209415

30. K.T. Sweeney, T.E. Ward, S.F. McLoone, Artifact removal in physiological signals-practices and possibilities. IEEE Trans. Inf Technol. Biomed. **16**(3), 488–500 (2012)

31. D.A. Kaiser, Basic principles of quantitative eeg. J. Adult Dev. **12**(2), 99–104 (2005). https://doi.org/10.1007/s10804-005-7025-9

32. L. Meirovitch, *Fundamentals of Vibrations*, ser. McGraw-Hill Higher Education (McGraw-Hill, 2001). https://books.google.com/books?id=u358QgAACAAJ

33. B.N. Cuffin, D. Cohen, Comparison of the magnetoencephalogram and electroencephalogram. Electroencephalogr. Clin. Neurophysiol. **47**(2), 132–146 (1979). https://www.sciencedirect.com/science/article/pii/0013469479902153

34. A.S.C.T. Staff, A. Gelb, A.S. Corporation, *Applied Optimal Estimation*, ser. Mit Press (MIT Press, 1974). https://books.google.com/books?id=KlFrn8lpPP0C
35. M. Sun, D. Zeng, Y. Wang, Modelling temporal biomarkers with semiparametric nonlinear dynamical systems. Biometrika **108**(1), 199–214 (2020). https://doi.org/10.1093/biomet/asaa042
36. J.R. Petrella, W. Hao, A. Rao, P.M. Doraiswamy, Computational causal modeling of the dynamic biomarker cascade in alzheimer's disease. Comput. Math. Methods Med. **2019** (2019)
37. A. Root, Mathematical modeling of the challenge to detect pancreatic adenocarcinoma early with biomarkers. Challenges **10**(1) (2019). https://www.mdpi.com/2078-1547/10/1/26
38. P. Van Overschee, B. De Moor, *Subspace Identification for Linear Systems: Theory-Implementation-Applications* (Springer Science & Business Media, Berlin, 2012)
39. T. Kato, *Perturbation Theory for Linear Operators*, vol. 132 (Springer Science & Business Media, Berlin, 2013)
40. J. Roubal, P. Husek, J. Stecha, Linearization: students forget the operating point. IEEE Trans. Educ. **53**(3), 413–418 (2010)
41. C.-T. Chen, B. Shafai, *Linear System Theory and Design*, vol. 3 (Oxford University Press, New York, 1999)
42. P. Antsaklis, A. Michel, *A Linear Systems Primer* (Birkhäuser, Boston, 2007). https://books.google.com/books?id=7W4Rbqw_8vYC
43. C.M. Sweeney-Reed, S.J. Nasuto, M.F. Vieira, A.O. Andrade, Empirical mode decomposition and its extensions applied to eeg analysis: a review. Adv. Data Sci. Adapt. Anal. **10**(02), 1840001 (2018). https://doi.org/10.1142/S2424922X18400016
44. J.J. Newson, T.C. Thiagarajan, Eeg frequency bands in psychiatric disorders: a review of resting state studies. Front. Hum. Neurosci. **12**, 521 (2019). https://www.frontiersin.org/article/10.3389/fnhum.2018.00521
45. R. Wang, Z.-Z. Zhang, J. Ma, Y. Yang, P. Lin, Y. Wu, Spectral properties of the temporal evolution of brain network structure. Chaos: Interdiscip. J. Nonlinear Sci. **25**(12), 123112 (2015). https://doi.org/10.1063/1.4937451

System Identification of Brain Wave Modes Using EEG

3

3.1 Introduction and Motivation

Rigorous descriptions for both the macro and microscopic intricacies of the brain are a cross-disciplinary scientific challenge. There is potential in this challenge for an improved understanding of the interaction between our physiology and the stimuli in our environment. This is especially important when considering that our environments increasingly contain both physical and digital objects in coexistence. Consider that the internet of things (IoT), brain-computer interfaces (BCI), and collaborative robots (co-robots) are just some of the developing technologies connecting our carbon-based brain to silicon-based processors. In each of these cases, there is necessarily a transfer of information beginning at the cellular level of the brain and ending with some digital interpretation of the information. Encoding this physiological information toward digital interpretability is a key component of the search for greater performance and safety [1]. Therefore, there is a need for rigorous, transparent descriptions of the dynamics of human physiology as it is relevant to our increasingly digital environments. Here, we present such a description for the macroscopic dynamics of neural activity in the form of brain wave modal analysis.

3.1.1 Motivation: Dynamical Models of Biomarkers

The description of human cognitive function as a dynamic process with static architecture is well established [2]. In mechanical systems, which are often also dynamic with static architecture, it is common practice to use the observable outputs of a system to form descriptions of the system's dynamics [3]. There is a great deal of existing work on the identification and analysis of complex system dynamics including flapping wing robots [4, 5], insect flight control [6], and human electrodermal activity [7].

© The Author(s), under exclusive license to Springer Nature Switzerland AG 2023

T. D. Griffith et al., *A Modal Approach to the Space-Time Dynamics of Cognitive Biomarkers*, Synthesis Lectures on Biomedical Engineering, https://doi.org/10.1007/978-3-031-23529-0_3

We focus on considering the spatial and temporal dynamics of brain waves jointly. While EEG measures have millisecond temporal resolution, a limited number of electrodes can be placed on the head. For this reason, blood-oxygen-level-dependent (BOLD) functional magnetic resonance imaging (fMRI) is often employed instead of EEG when analyzing specific anatomical regions of the brain. fMRI measures have higher spatial resolution than EEG measures but rarely collect scans faster than 1 Hz [8], which can lead to problematic statistical analysis [9]. As a result, much effort has been dedicated to jointly analyzing the spatial information in BOLD fMRI measures and the temporal information in EEG measures [10–12]. Further, current MRI machines are difficult to integrate with many experiments because of their size and cost [13]. If improved spatial information can be extracted from EEG measures and integrated with the inherent temporal dynamics, it may be possible to correlate EEG with the subsequent physiological BOLD changes.

Spatio-temporal brain wave dynamics contain structured linear and nonlinear behavior, as well as unstructured random noise. While EEG measures can be nonstationary, the use of highly derived nonlinear indices has repeatedly been shown to be less effective than robust linear methods [14]. Such highly derived models of EEG dynamics are often difficult to interpret and analyze [15]. It is for this reason that the vast majority of scientific and clinical conclusions about EEG recordings come from relatively simple, robust measures, such as analysis of the raw time series [16], the power spectrum [17], and the independent [18] or principle [19] components of EEG data. At their core, even most analyses from fMRI recordings are also linear [20].

This well-known phenomenon, where the system dynamics are nonlinear but the most effective analytic tools are linear, is common in other engineering systems. Dynamic systems on earth and in space are frequently subject to nonlinear effects, which may be very strong, but it again has repeatedly been shown that linear formulations are sufficient to steer autonomous ground robots [21], guide commercial aircraft [22], and even land on the moon [23]. Now, we do not mean to imply that the brain is exactly like these other systems, but rather that linear dynamical systems are appropriate to the study of *EEG measures* [3]. In this work, we seek to describe a rigorous, canonical engineering approach to the modeling of EEG dynamics through the lens of linear state-space modeling. This approach is comprised of four parts:

1. identification of an LTI model with modern system identification tools;
2. validation of the identified model;
3. coordinate transformation of the identified model to generate highly interpretable spatio-temporal modes; and
4. analysis of the extracted modes.

In the following section, we will describe in detail each of these steps. While we have targeted this application toward EEG recordings, the analysis is agnostic to the data in the output measure and may be readily applied to other time series biomarker data.

3.1.2 Linearization of Neural Dynamics

Almost all systems exhibit nonlinear behavior. Much of modern control theory is dedicated to analyzing and stabilizing nonlinear systems. Instead of viewing (2.2) as a global model for the brain at all times, we recognize there are nonlinearities in the dynamics and treat the modes as a model of the locally linear dynamics around a stable operating point. There are many examples of successful nonlinear plant control using linearized models about an operating point [24–26]. The linearization is as important as the model. The operating point x_i, y_i informs the state of the dynamical system in addition to the extracted modes. By tracking the model error $e_y = \hat{y} - y$, we can move the linearization point as needed by recalculating the modal decomposition.

We note that one of the benefits of linearization is the ability to run estimators and evaluate modal results in near real time. We make the distinction near real time because the determination of modes around an operating point can be computationally intensive when long time series are considered. Once this initial linearization and subsequent decomposition is completed, a state-space model in (2.5) is identified where $\hat{y}(k)$ is our best recreation of the measured EEG signal. Accordingly, the whole literature of optimal state estimation becomes available to us, and we can track the modal dynamics in real time while keeping the error between the estimated output limited to some radius centered at zero. The "features" v_i are generated in a canonical fashion and are directly tied to the measured data y through least squares projections and linear transformations only. Any result from the state-space model can be directly tied back to the data.

3.1.3 Evaluation of System Identification Techniques for Brain Wave Modeling

In this work, the focus is on the formulation and analysis of the *linear modal patterns* in EEG brain wave recordings. As mentioned in Sect. 2.2, while these linear state-space models represent a very simplified view of brain wave dynamics, they contain useful diagnostic information. The synthesization of state-space models from a given plant's data falls under the broad engineering topic of system identification.

In the study of brain wave dynamics, we may only measure the output of the EEG device, which is denoted here as $y(t)$. In many other applications, knowledge of the system input $u(t)$ is also available. The vast majority of system identification techniques rely on analysis of the relationship between input and output measurements, typically through linear systems analysis and least squares regression [27]. These include transfer function estimation [28], Autoregressive-moving-average estimation (ARMAX) [29], and unstructured or parameterized state-space estimation [30], among others. Here, we are restricted to the use of output-only system identification techniques, such as numerical algorithms for subspace state-space system identification (N4SID) [31], frequency domain decomposition (FDD)

[32], dynamic mode decomposition (DMD) [33], and output-only modal analysis (OMA) [34]. As we will see, these algorithms generally leverage observability concepts to formulate minimum variance estimates of the matrix pair (A, C). While the outcome is common, the variety of algorithmic approaches here means it is not immediately obvious which method will yield better results than another. As a result, the discussion that follows covers our approach to evaluating a number of output-only system identification techniques against one another, in order to develop a systematized modeling procedure for the analysis of brain wave dynamics. First, we take a brief survey of the considered algorithms.

3.1.4 Overview of Considered Output-Only Algorithms

The following output-only system identification algorithms were considered in this work. The methods listed here were selected for their robustness to process noise, computational tractability, and widespread use among engineering practitioners.

1. **N4SID:** The N4SID algorithm is ubiquitous for its inception as one of the earlier successful methods for synthesizing non-polynomial models. It is particularly well suited for multi-input multi-output (MIMO) systems, and has formulations as both an input-output and output-only approach. Since it is an explicit calculation, requiring no optimization algorithm, the common issues of convergence and initial condition sensitivity are avoided. The N4SID approach assumes the system dynamics follow the following discrete-time model analogous to our continuous formulation in (2.2)

$$x_{k+1} = Ax_k + Bu_k + v_x \tag{3.1}$$

$$y_k = Cx_k + Du_k + v_y, \tag{3.2}$$

or when the input is unknown

$$x_{k+1} = Ax_k + v_x \tag{3.3}$$

$$y_k = Cx_k + v_y. \tag{3.4}$$

Leaving the proof and other technical details aside (and referring the reader to [31]), we state for the time being that the deterministic portion of the dynamics may be identified as long as the dynamics are stabilized and all states are observable. The formulation of the plant (A, B, C) with input-output data, or (A, C) with the output data only, follows readily from a least squares projection of the Kalman states, which is a minimum variance estimate. This is a promising approach for brain wave modeling since it has a little restriction on plant parametrization, is well suited for multiple output dynamics, and can be used with output data alone. Of course, in the output-only case, the effect of the input is ignored and may leak into the formulation of the system dynamics if the input is not stochastic in nature.

2. **NExT ERA:** The natural excitation eigensystem realization algorithm (NExT ERA) [35] is a popular output-only system identification approach in the analysis of civil and aerospace structures because it operates in modal space, which is well studied and physically significant for most structures. Unlike the N4SID approach, NExT ERA assumes a completely deterministic plant of the form

$$\dot{x} = Ax + Bu \tag{3.5}$$

$$y = Cx \tag{3.6}$$

is to be identified. As in N4SID, if there is no input data, the formulation is simplified to

$$\dot{x} = Ax \tag{3.7}$$

$$y = Cx. \tag{3.8}$$

As with many system identification methods, NeXT ERA makes use of the generalized Hankel matrix. If we had N discrete recordings of a system output vector y, the data may be represented in the block matrix

$$Y = \begin{bmatrix} | & | & | & | \\ y_1 & y_2 & \cdots & y_N \\ | & | & | & | \end{bmatrix}. \tag{3.9}$$

When formulating state-space models from output data, it is often useful to cycle the recorded data and perform individual calculations on the sequential blocks of data. The generalized Hankel matrix H is a collection of all the shifted block matrices of data

$$H = \begin{bmatrix} Y_{(1:N-2s)} \\ Y_{(2:N-2s+1)} \\ \vdots \\ Y_{(2s:N)} \end{bmatrix} = \begin{bmatrix} H_p \\ H_f \end{bmatrix}. \tag{3.10}$$

The number of block rows in the Hankel matrix is $2s$, which is often termed the total data shift. Many system identification algorithms treat the top half of the Hankel matrix rows as "past" data H_p, and the bottom half of the rows as "future" data H_f. The identification task is then to determine a minimum variance system that advances the "past" data into the "future" data through various projections. In the case of NeXT ERA, a singular value decomposition (SVD) of the Hankel matrix is performed on the past data

$$H_p = USV^*, \tag{3.11}$$

and the system realization of order n immediately follows from the SVD and the future data

$$A = S_n^{1/2} U_n^* H_f V_n S_n^{-1/2} \tag{3.12}$$

$$B = S_n^{1/2} V_n^T (:, p) \tag{3.13}$$

$$C = U_n(q, :) S_n^{1/2} \tag{3.14}$$

where q and p represent the necessary row or column truncation for the operation. These matrices may then be modally decomposed as introduced in Sect. 2.3.1 for a modal view of the measured dynamics. While this is a promising algorithm because it is focused on a spatio-temporal view of the system dynamics and leverages the eigenvalue problem, NeXT ERA does not account for stochastic effects or modeling uncertainties directly.

3. **OMA** Output-only modal analysis (OMA) is included as the stochastic subspace identification method of interest to this work [36]. Conceptually, it is similar to the operating principle of NeXT ERA, where an oblique projection from past to future data leads to a formulation of the system matrices. Crucially, OMA takes a stochastic view of the dynamics in discrete time

$$x_{k+1} = Ax_k + v_x \tag{3.15}$$

$$y_k = Cx_k + v_y, \tag{3.16}$$

where v_x and v_y are zero mean, white vector sequences. This is a slightly more restrictive condition that we posed in (2.4). It's also worth noting that by stochastics here, we simply mean the dynamics have some randomness associated with their evolution, not that they are necessarily completely random. Clearly, by the form of (3.15), there is still analytically valuable content in the pair (A, C). In particular, we should view v_x here as a random, unknown input to the otherwise deterministic formulation of the system matrices. That in mind, the oblique projection from past states to future states in OMA takes the form

$$O = \mathbb{E}(H_f | H_p) = \begin{bmatrix} o_1 & o_2 & \dots & o_s \end{bmatrix}, \tag{3.17}$$

where $\mathbb{E}$ is the expectation. Van Overschee and De Moor [36] demonstrated that O can be directly calculated from the generalized Hankel matrix with

$$O = H_f H_p^* (H_p H_p^*)^{-1} H_p. \tag{3.18}$$

Conceptually, this matrix consists of the system free decays at the different "initial" conditions as seen in H_p. Using the matrix exponential, the columns of O may be expressed through the observability matrix

$$O = \Gamma_s X_0 \tag{3.19}$$

$$= \begin{bmatrix} C \\ CA \\ CA^2 \\ \vdots \\ CA^{s-1} \end{bmatrix} X_0. \tag{3.20}$$

If all the initial (Kalman) states were immediately available, we could simply invert X_0 and solve for the system matrices (A, C). Instead, we estimate Γ and X_0 with the SVD

$$\hat{\Gamma} = U S^{1/2} \tag{3.21}$$

$$\hat{X}_0 = S^{1/2} V^*. \tag{3.22}$$

Obviously, this formulation, and output-only system identification result generally, will not be unique. However, it gives us a method to readily obtain an estimate of (A, C). C is simply the first block of $\hat{\Gamma}$

$$C = \hat{\Gamma}_{1:1}, \tag{3.23}$$

while A can be derived from the second block and C^{-1} or the top and bottom blocks

$$\hat{\Gamma}_{2:s} A = \hat{\Gamma}_{1:s-1}. \tag{3.24}$$

From here, the modal parameters of the system may be readily calculated through the appropriate transformation matrix [37].

One of the particular advantages of OMA, in this case, is the assumption of zero mean white noise in the driving states. While EEG recordings are known to be nonstationary, they are generally processed in a way that makes them zero mean. That is, $\mathbb{E}(y) = 0$. This is often not the case for traditional dynamical systems, where non-zero mean speed, acceleration, and force must be considered.

4. **DMD** Finally, dynamic mode decomposition (DMD) was evaluated [33]. DMD is a fairly modern data driven state-space identification algorithm first developed for fluid dynamic applications. Accordingly, it is especially oriented for the generation of physically significant spatial parametrizations, which is why it was selected for use in the analysis of brain waves. Unlike the other methods presented thus far, it assumes complete access to the locally linear dynamics of the plant

$$\dot{x} = Ax. \tag{3.25}$$

The identification scheme is to determine the matrix A which advances $x(t)$ to $x(t+1)$ as in

$$\min ||x(t+1) - Ax(t)||_2. \tag{3.26}$$

There are theoretical and algorithmic extensions to this approach which allow for the consideration of multiple recordings $\{x(t+2), x(t+3), \ldots, x(n))$, multiple time scales, and reduced system orders, but we leave those to [33]. While its computational complexity is very low, it is sensitive to transient time phenomena and is generally more suited to deterministic systems of high cardinality, rather than moderate dimension deterministic-stochastic systems like brain wave dynamics.

Overall, we must remember that models generated through system identification algorithms such as those discussed here are only as good as their validation. In particular, all of these methods rely on the SVD, which is known to struggle with data invariances (e.g. low rank rotation). Additionally, none of these algorithms generate unique solutions except DMD. This is the result of considering the output matrix C. Conceptually, A and C are tied together through the dynamic process, so any change in A could potentially be canceled out by the requisite change in C. In the case of DMD, by assuming the entire plant state is available, we lose the ability to identify potentially important internal dynamics, especially when $\dim(y) << \dim(x)$. Much of the existing literature involves the comparison of at least two system identification approaches for a given problem [38–40]. In many cases, mathematical equivalence is even possible with the right assumptions or initial conditions [36]. Models formulated this way can only model the dynamics in a given set of measurements. For example, if a system has a destabilizing natural frequency, and that frequency is not captured in the Hankel matrix, the model (and therefore any controller) will be blind to that instability. That is, *you cannot model what you do not excite*. While this is important for physical systems, it is not a concern for EEG measures, which do not diverge. In fact, it may even be a benefit, because we can compare the range of different models from one set of stimuli to another checking for differences in the extracted modes.

3.2   Evaluation of Output-Only System Identification Techniques

The techniques discussed above in Sect. 3.1.4 above were selected on their structure as output-only system identification techniques with a specific focus on modal analysis. As mentioned, many of these algorithms can be shown as identical under the right constraints, but this does not mean that the resultant models will be equivalent [30]. In this section, we will discuss the analysis we performed to compare the algorithms for the initial modeling of EEG brain waves.

3.2.1 Model Variance Among Subjects

Model validation is a critical step in system identification. An improperly validated model may obscure potential destructive failure modes, especially if these modes were not excited in the data used to create the model. No system identification technique is omniscient; they cannot model what they do not have information to model. Therefore, it is crucial to validate models against an outcome of interest. Frequently, model validation occurs through a comparison of model parameters with expected results from first principles, CFD, or experimental data.

Here, we are in a unique situation with EEG data for two reasons. First, we are not attempting to do any kind of feedback control. When closing a control loop around a system identification model, there is a risk of exciting unstable eigenvalues in the true system plant. Since we are not interested in any form of control feedback for brain waves, this is not an issue. Second, and more importantly, EEG brain waves do not diverge, so we need not worry about collecting sufficient data to model instabilities. This is all to reinforce a key argument in our approach. The use of output-only system identification serves as a tool for the analysis of brain waves. We do not seek to identify a global model for brain waves, but rather to identify the dynamics which are actively present in a given EEG brain wave recording.

Of course, this makes our evaluation of individual models more challenging. There is not a clear control outcome that we can evaluate the performance of to say one model is better than another. Further, there are no first principles or "wind tunnel" data against which we can compare the synthesized dynamics. Therefore, our initial approach was to select system identification techniques that consistently identified similar models for the same person. We hypothesize, and will show in [41], that data from the same person should have shared modes, because brain waves are known to be interpersonally dependent. So too should our modeling approach identify these common modes. Much of the analysis in this work, including this initial evaluation of system identification techniques, relies on the Dataset for Emotion Analysis using EEG, Physiological, and Video Signals.

3.2.2 Databases for Initial Brain Wave Modeling

Dataset for Emotion Analysis using EEG, Physiological, and Video Signal: The Database for Emotion Analysis using Physiological Signals [42] was selected for demonstration due to the high number of subjects in the dataset and the nature of the elicitation in the experimental methods. The database contains 40 distinct minute long trial videos for each of 32 different participants who watched a variety of music videos. It employs 32 EEG channels, placed according to the 10–20 system, and is available in a preprocessed format at 128Hz. Also included are self-reported ratings of experienced emotions according to the Valence-Arousal-Dominance model.

In this work, we make no additional preprocessing modifications to the data. Our goal is to model the measured output y, so we wish not to modify the original data too much. In particular, we believe many of the common EEG artifacts are endemic to the brain wave system. Heart, eye, and muscle activity stem first from brain activity, so our focus is on modeling the EEG measurement.

3.2.3　Joint Distributions of Modal Parameters

In order to determine which system identification technique yields consistent eigenmodes for a given single subject, we analyzed the distribution of the real and imaginary parts of all the eigenmodes for all subjects in the DEAP database across system identification algorithms. We set aside the exact modeling details for the individual algorithms, as for this comparative evaluation the results should be valid as long as the parameters (system order, number of Hankel rows, etc) are the same for each method. Further, a great number of references cover the considerations and application of these methods [30, 33, 41, 43]. Moreover, the specific initialization and synthesization of models from each of these algorithms greatly depend on the number of recording channels, length of the recording, sampling frequency, and other modeling considerations. What is important is that each of the algorithms generates the matrix pair (A, C) which has associated with it eigenvectors and eigenvalues. The eigenvectors $v \in \mathbb{C}^n$, which is to say they are complex and so have real and imaginary components. Figure 3.1 shows the distribution of the real and imaginary parts of each eigenvector identified by each system identification technique for the subjects in the DEAP database. Note, that the distributions of both the OMA and DMD algorithms are most closely grouped. For example, the OMA eigenvectors have a minimum value of negative three and a maximum of four, while the NeXT eigenvectors range from -100 to 80. Therefore, we selected the OMA and DMD algorithms for further analysis. This was merely an initial attempt to provide some form of rigorous comparison between the algorithms. Again, because we wish to analyze the eigenvectors for spatio-temporal patterns, it is important to have consistent results from one subject to the next. If one set of eigenvectors is an order of magnitude greater than another set from the same system identification routine, it will be difficult to compare them. Rather, we should select the algorithms which more closely group their models.

3.2.4　Assumptions and Constraints

Before proceeding to the technical details of OMA and DMD and their application to the DEAP database, it is worth a brief discussion of the assumptions and constraints of each. Understanding how the strengths and weaknesses of each method fit into the analysis of EEG brain waves is important contextually.

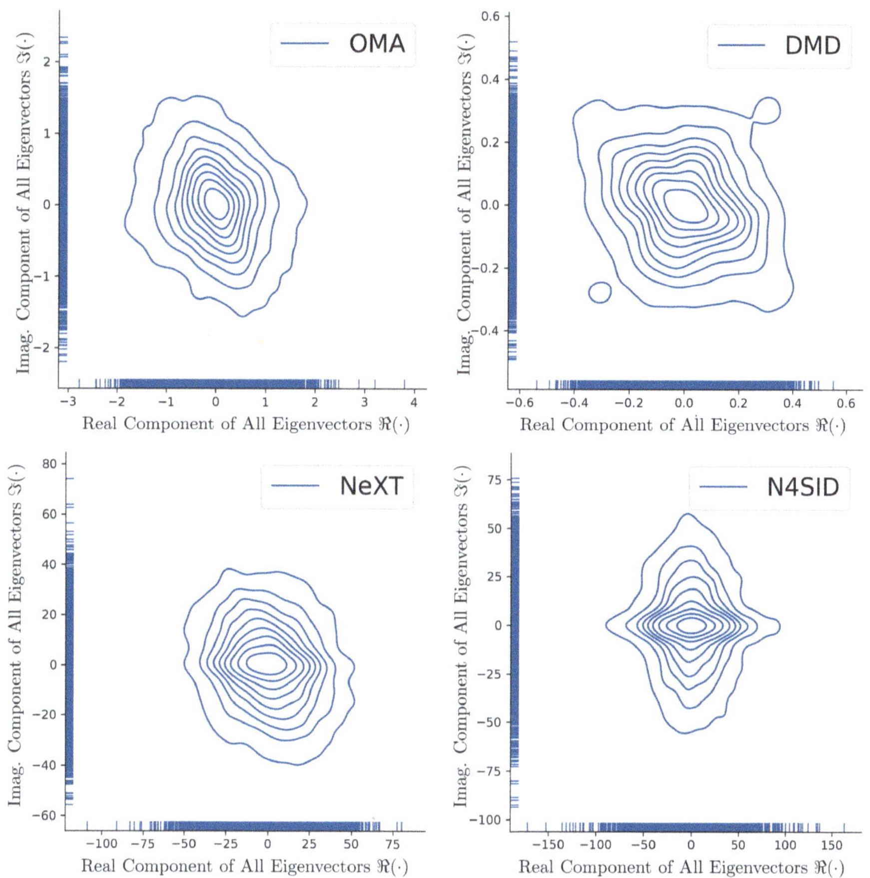

Fig. 3.1 Distributions of identified eigenvectors for all subjects in the DEAP database. Notice, by their axes, the OMA and DMD algorithms have the most consistent model synthesis (e.g. NeXT modes have real components ranging from -100 to 75, while OMA modes have real components ranging from -3 to 4)

- **OMA:** It is not surprising that OMA is particularly well suited to this problem. First, OMA has rigorous theoretical roots in the analysis of noisy systems under uncertain loading conditions (e.g. earthquakes [44]). While NeXT and N4SID can accommodate a known input in their analysis, OMA is strictly targeted at systems where the input is unknown. OMA uniquely makes no restrictions on the nature of the true input and does not require knowledge of the system boundary conditions. While some references treat the white noise term v_x as an input, we prefer to conceptualize it as process noise in alignment with our formulation in Sect. 2.2. Two significant assumptions accompany the use of this algorithm. First, the models OMA identifies are LTI. If the linear dynamics change as a function of time, or there are significant nonlinear effects, the modal decomposition will likely be invalid. We treat this issue by recalculating the modes about a linearized cognitive state, but it would be equally valid to select a linear parameter varying (LPV) or nonlinear system identification method instead. Second, the system is assumed to be observable. This is a nontrivial assumption for EEG brain waves. Because of the interference introduced by the skull and the CSF, there is no obvious reason to expect that all the relevant dynamics will be available in any output measure y.

- **DMD:** The strength of DMD lies in the simplicity of its formulation. Because of this, it is computationally very tractable, which makes testing new modeling hypotheses quick. Further, it is closely related to Koopman analysis and may be extended to treat nonlinear parameters. However, it does not leverage any form of stochastic analysis like the other three algorithms here. The dynamics are assumed to be completely observable, and there is no robustness to general process disturbances. As a result, DMD is a measure that performs best in systems with high-density spatio-temporal measurements which is not necessarily the case with EEG data.

While DMD shows a consistent decomposition of the data for a given subject, we are especially interested in methods that allow for the reconstruction of internal dynamics. Since DMD assumes the entire state is measurable, it will likely perform poorly with a reduced sensing suite (e.g. fewer EEG electrodes). Accordingly, we especially recommend the OMA algorithm for system identification of biomarkers. We now turn to a brief demonstration of the modeling output from OMA on EEG data.

3.3   Results

In this section, we will discuss the resultant output of our proposed modeling approach. As a reminder, each of the discussed system identification routines synthesizes a model in the form of (2.5) which may be represented in a modal form through a coordinate transform. The resultant modes describe the measured dynamics with a spatial pattern at an oscillating frequency, together forming an eigenmode. Figure 3.2 shows a collection of some of the eigenmodes obtained from a representative subject in the DEAP database. The displayed modes were selected as those with the most dynamic mode shape. Again, a superposition of all the modes precisely recreates the observed EEG measure, and the state-space formulation of the model is amenable to a wide variety of state estimators, such as the Kalman filter. Figure 3.3 shows the true EEG data against the superposition of the eigenmodes for the first ten EEG channels. Notice that the modes reconstruct the original signal well.

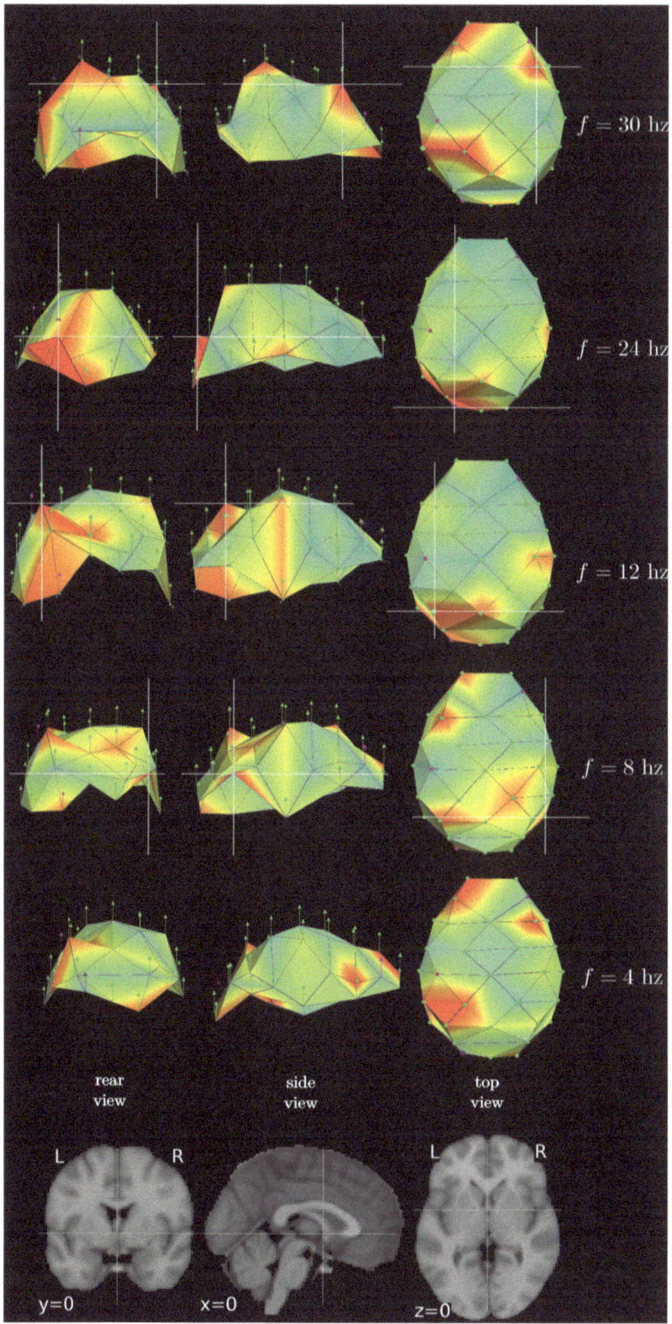

Fig. 3.2 A selection of demonstrative eigenmodes from a single subject in the DEAP database

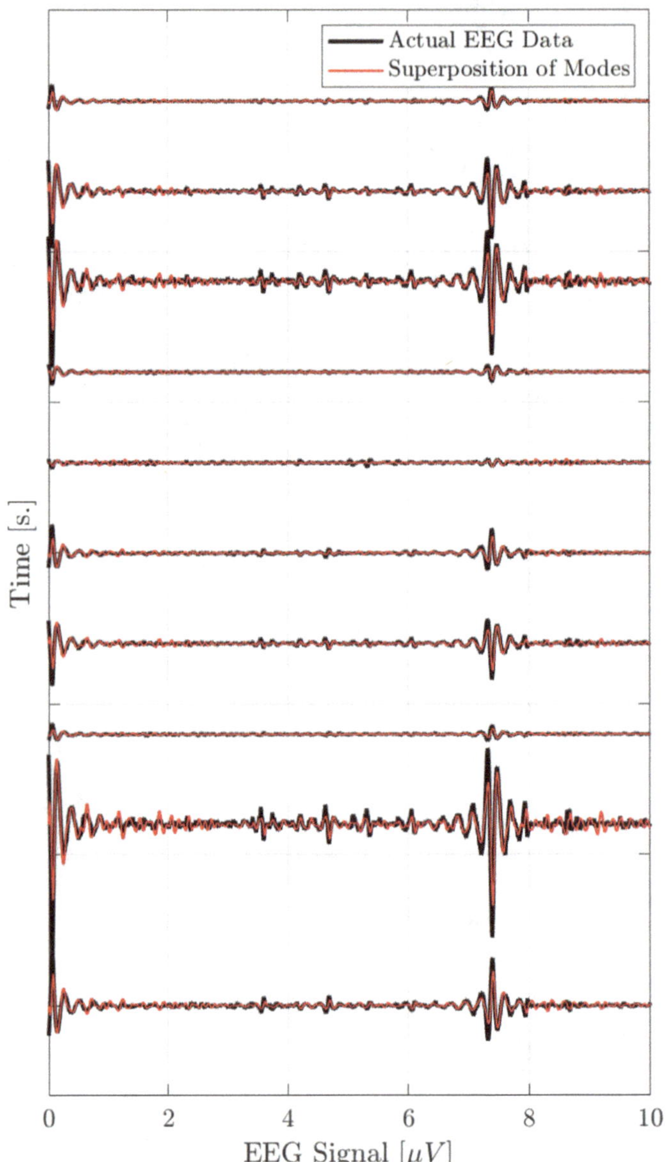

Fig. 3.3 A superposition of identified modes recreates the EEG signal with moderate fidelity. The mean absolute percentage error (MAPE) is 5.5% for this trial

While these shapes were derived from a surface measure on the scalp, they yield a very natural 3D representation of the measured EEG wave. Since the spatial dependence of brain waves is well studied, we believe this approach is well suited to real time analysis of the spatio-temporal dynamics observed in EEG waves. Once the pair (A_m, C) has been extracted, the tracking in Fig. 3.3 is immediately available in real time. The authors strongly encourage readers to view the animated modes at https://tdgriffith.github.io/EEG-imaging/ as they especially demonstrate the physical significance of the modes. Again we wish to draw attention to the symmetry and significance of Figs. 3.2 and 3.3 together. The original brain wave measurement occurred in "EEG space", and through this elegant approach, we extracted precise spatio-temporal dynamics in state-space that superpose to recreate the original measure. Much has been made of the high temporal frequency of EEG data, with the accepted drawback of limited spatial density. In the same way that there is a real need for faster fMRI measures [45], there is a need for EEG measures with better spatial resolution. This approach offers a solution that balances the need for biomarkers that are feasible in driving simulators, marketing studies, and clinical trials with the need for the recovery of spatial dynamics.

3.3.1 Reducing the Number of EEG Channels

A particularly elegant application of this approach is the consideration of observability as a metric for reducing the number of EEG channels. The DEAP database was recorded using a 60 channel EEG device, which is a relatively high spatial resolution for EEG measures. Any model in the form of (2.5) will employ a C matrix with 60 rows. If we wish to validate our models in a separate experiment, it may be cumbersome to use the same 60 channel EEG device. As long as the observability of the pair (A_m, C) is preserved, any outcome based on the eigenmodes of A_m can be reconstructed. That is, observability provides an analytical tool to verify that a reduced number of sensors can still measure a given set of eigenmodes in real time. Figure 3.4 shows the minimum number of rows in C that preserve observability as a function of how many eigenmodes are modeled for all subjects in the DEAP database. As the number of modes increases, the number of sensors needed also increases, but at a much slower rate. Note that no more than four sensors are required to observe as many as 120 modes in real time. Observability provides both the number of channels needed to measure the eigenmodes and the spatial location of the channels. If a metric is developed using the eigenmodes from A_m, we can quickly determine the appropriate and minimal number of sensors needed to recreate that metric through observability.

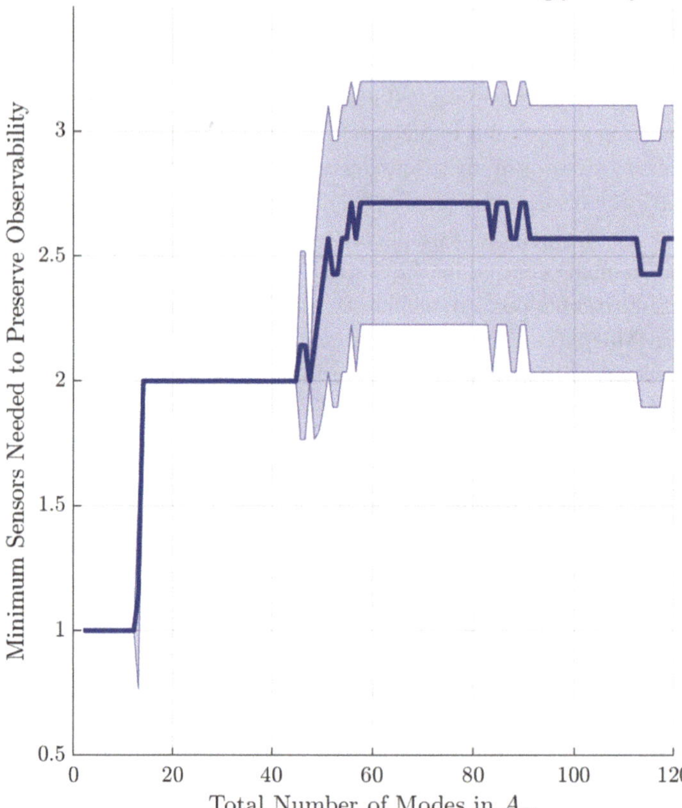

Fig. 3.4 Number of sensors (rows in C) needed to preserve observability as the number of extracted modes is increased. Note that no more than 4 sensors are needed to observe as many as 120 modes

3.4   Conclusions

In this work, we demonstrated that modern system identification routines are appropriate for the modeling of biomarkers, especially EEG measures. By leveraging the large body of knowledge associated with modern dynamical systems, we elegantly treat the need for spatio-temporal EEG dynamics while handling the noisy nature of EEG data. The modal decomposition of state-space models is particularly important since it casts a nondescript coupled ODE into a form that may easily be understood as a discrete collection of eigenmodes at different frequencies. The eigenmodes offer a robust feature extraction technique for the analysis of EEG waves, which we will explore in future work.

3.4.1   Recommendations

The modal view offers improved insight into the brain's function connectivity. Crucially, the imaging method presented in this work decomposes EEG data in near real time and has both spatial and temporal dynamics jointly. We hypothesize that individual modes are connected to human cognition. More work is needed to make this claim rigorously, but successful identification of the connection between modes and cognition would suggest new clinical applications for mode targeting [46]. Crucially, others have achieved experimental success with modes. In [47], authors achieved a substantial reduction of beta amyloid plaques, a frequent indicator of Alzheimer's disease, in the visual cortex of mice by flickering LED lights at 40 Hz. This approach was extended recently to demonstrate measurable improvements to memory and cognition [48].

In our view of EEG dynamics, it is not surprising that mode targeting resulted in improved cognitive performance since the modes have physical significance in the brain. By applying specific stimuli, individual modes can be exited. More work is needed to connect specific external stimuli (light, sound, and vibration) to EEG modes, but these existing results suggest that other neurological diseases may be diagnosed and even treated with the modal view.

With an overview of the system identification techniques for linear models of brain wave dynamics complete, we turn to a more detailed analysis of the OMA and DMD algorithms in Chap. 4. This includes technical modeling considerations and some analysis of the brain wave modes.

References

1. M.R. Endsley, From here to autonomy: lessons learned from human–automation research. Hum. Fact. **59**(1), 5–27 (2017). pMID: 28146676. https://doi.org/10.1177/0018720816681350
2. E. Bullmore, O. Sporns, Complex brain networks: graph theoretical analysis of structural and functional systems. Nat. Rev. Neurosci. **10**(3), 186–198 (2009)
3. J. Hespanha, *Linear Systems Theory: Second Edition* (Princeton University Press, Princeton, 2018). https://books.google.com/books?id=eDpDDwAAQBAJ
4. C. Altenbuchner, J.E. Hubbard Jr., *Modern Flexible Multi-Body Dynamics Modeling Methodology for Flapping Wing Vehicles* (Academic, Cambridge, 2017)
5. E.A. Morelli, J.A. Grauer, Practical aspects of frequency-domain approaches for aircraft system identification. J. Aircr. **57**(2), 268–291 (2020)
6. A.L. Stöckl, K. Kihlström, S. Chandler, S. Sponberg, Comparative system identification of flower tracking performance in three hawkmoth species reveals adaptations for dim light vision. Philos. Trans. R. Soc. B: Biol. Sci. **372**(1717), 20160078 (2017). https://royalsocietypublishing.org/doi/abs/10.1098/rstb.2016.0078
7. M.R. Amin, R.T. Faghih, Sparse deconvolution of electrodermal activity via continuous-time system identification. IEEE Trans. Biomed. Eng. **66**(9), 2585–2595 (2019)
8. M.J. Sturzbecher, D.B. de Araujo, Simultaneous eeg-fmri: integrating spatial and temporal resolution, in *The Relevance of the Time Domain to Neural Network Models* (Springer, Berlin, 2012), pp. 199–217

9. B.O. Turner, E.J. Paul, M.B. Miller, A.K. Barbey, Small sample sizes reduce the replicability of task-based fmri studies. Commun. Biol. **1**(1), 1–10 (2018)

10. M.A. Pisauro, E. Fouragnan, C. Retzler, M.G. Philiastides, Neural correlates of evidence accumulation during value-based decisions revealed via simultaneous eeg-fmri. Nat. Commun. **8**(1), 1–9 (2017)

11. M. Bullock, G.D. Jackson, D.F. Abbott, Artifact reduction in simultaneous eeg-fmri: a systematic review of methods and contemporary usage. Front. Neurol. **12**, 193 (2021). https://www.frontiersin.org/article/10.3389/fneur.2021.622719

12. Y. He, M. Steines, J. Sommer, H. Gebhardt, A. Nagels, G. Sammer, T. Kircher, B. Straube, Spatial-temporal dynamics of gesture-speech integration: a simultaneous eeg-fmri study. Brain Struct. Funct. **223**(7), 3073–3089 (2018)

13. L.L. Wald, P.C. McDaniel, T. Witzel, J.P. Stockmann, C.Z. Cooley, Low-cost and portable mri. J. Magn. Reson. Imaging **52**(3), 686–696 (2020). https://onlinelibrary.wiley.com/doi/abs/10.1002/jmri.26942

14. D.A. Kaiser, Basic principles of quantitative eeg. J. Adult Dev. **12**(2–3), 99–104 (2005)

15. L.C. Parra, C.D. Spence, A.D. Gerson, P. Sajda, Recipes for the linear analysis of eeg. Neuroimage **28**(2), 326–341 (2005)

16. N. Jadeja, *How to Read an EEG* (Cambridge University Press, Cambridge, 2021). https://books.google.com/books?id=GskwEAAAQBAJ

17. S. Liu, J. Shen, Y. Li, J. Wang, J. Wang, J. Xu, Q. Wang, R. Chen, Eeg power spectral analysis of abnormal cortical activations during rem/nrem sleep in obstructive sleep apnea. Front. Neurol. **12** (2021). https://www.frontiersin.org/article/10.3389/fneur.2021.643855

18. M. Klug, K. Gramann, Identifying key factors for improving ica-based decomposition of eeg data in mobile and stationary experiments. Eur. J. Neurosci. **54**(12), 8406–8420 (2021). https://onlinelibrary.wiley.com/doi/abs/10.1111/ejn.14992

19. R.J. Barry, F.M. De Blasio, Eeg frequency pca in eeg-erp dynamics. Psychophysiology **55**(5), e13042 (2018). https://onlinelibrary.wiley.com/doi/abs/10.1111/psyp.13042

20. J.-B. Poline, M. Brett, The general linear model and fmri: does love last forever? NeuroImage **62**(2), 871–880 (2012). 20 Years of fMRI. https://www.sciencedirect.com/science/article/pii/S1053811912001607

21. J. Hartzer, S. Saripalli, Vehicular teamwork: collaborative localization of autonomous vehicles (2021). arxiv:abs/2104.14106

22. A.U. Rehman, M.U. Khan, M.Z.H. Ali, M.S. Shah, M.F. Ullah, M. Ayub, Stability enhancement of commercial boeing aircraft with integration of pid controller, in *International Conference on Applied and Engineering Mathematics (ICAEM)* (2021), pp. 43–48

23. Y.-C. Ho, On centralized optimal control. IEEE Trans. Autom. Control **50**(4), 537–538 (2005)

24. J. Roubal, P. Husek, J. Stecha, Linearization: students forget the operating point. IEEE Trans. Educ. **53**(3), 413–418 (2010)

25. C.-T. Chen, B. Shafai, *Linear System Theory and Design*, vol. 3 (Oxford University Press, New York, 1999)

26. P. Antsaklis, A. Michel, *A Linear Systems Primer* (Birkhäuser, Boston, 2007). https://books.google.com/books?id=7W4Rbqw_8vYC

27. J.-N. Juang, *Applied System Identification* (Prentice-Hall Inc., Prentice, 1994)

28. H. Garnier, M. Mensler, A. Richard, Continuous-time model identification from sampled data: implementation issues and performance evaluation. Int. J. Control **76**(13), 1337–1357 (2003)

29. H.-F. Chen, L. Guo, Optimal adaptive control and consistent parameter estimates for armax model with quadratic cost. SIAM J. Control. Optim. **25**(4), 845–867 (1987)

30. L. Ljung, *System Identification: Theory for the User* (Pearson Education, London, 1998). https://books.google.com/books?id=fYSrk4wDKPsC

31. P. Van Overschee, B. De Moor, N4sid: subspace algorithms for the identification of combined deterministic-stochastic systems. Automatica **30**(1), 75–93 (1994). Special issue on statistical signal processing and control. https://www.sciencedirect.com/science/article/pii/0005109894902305

32. R. Brincker, L. Zhang, P. Andersen, Modal identification of output-only systems using frequency domain decomposition. Smart Mater. Struct. **10**(3), 441 (2001)

33. J.N. Kutz, S.L. Brunton, B.W. Brunton, J.L. Proctor, *Dynamic Mode Decomposition* (Society for Industrial and Applied Mathematics, Philadelphia, 2016). https://epubs.siam.org/doi/abs/10.1137/1.9781611974508

34. B. Peeters, G. De Roeck, Reference-based stochastic subspace identification for output-only modal analysis. Mech. Syst. Signal Process. **13**(6), 855–878 (1999)

35. R.S. Pappa, K.B. Elliott, A. Schenk, Consistent-mode indicator for the eigensystem realization algorithm. J. Guid. Control Dyn. **16**(5), 852–858 (1993). https://doi.org/10.2514/3.21092

36. P. Van Overschee, B. De Moor, *Subspace Identification for Linear Systems: Theory-Implementation-Applications* (Springer Science & Business Media, Berlin, 2012)

37. R. Brincker, P. Andersen, Understanding stochastic subspace identification, in *Conference Proceedings: IMAC-XXIV: A Conference & Exposition on Structural Dynamics* (Society for Experimental Mechanics, 2006)

38. S. Subramanian, G.B. Chidhambaram, S. Dhandapani, Modeling and validation of a four-tank system for level control process using black box and white box model approaches. IEEJ Trans. Electr. Electr. Eng. **16**(2), 282–294 (2021). https://onlinelibrary.wiley.com/doi/abs/10.1002/tee.23295

39. G. Sun, W. Li, Q. Luo, Q. Li, Modal identification of vibrating structures using singular value decomposition and nonlinear iteration based on high-speed digital image correlation. Thin-Walled Struct. **163**, 107377 (2021). https://www.sciencedirect.com/science/article/pii/S0263823120312428

40. N. Pandiya, W. Desmet, Direct estimation of residues from rational-fraction polynomials as a single-step modal identification approach. J Sound Vibr. **517**, 116530 (2022). https://www.sciencedirect.com/science/article/pii/S0022460X21005575

41. T.D. Griffith, J.E. Hubbard, System identification methods for dynamic models of brain activity. Biomed. Signal Proc. Control **68**, 102765 (2021). https://www.sciencedirect.com/science/article/pii/S1746809421003621

42. S. Koelstra, C. Muhl, M. Soleymani, J. Lee, A. Yazdani, T. Ebrahimi, T. Pun, A. Nijholt, I. Patras, Deap: A database for emotion analysis; using physiological signals. IEEE Trans. Affect. Comput. **3**(1), 18–31 (2012)

43. E.A. Morelli, V. Klein, *Aircraft System Identification: Theory and Practice*, vol. 2 (Sunflyte Enterprises Williamsburg, VA, 2016)

44. F. Pioldi, E. Rizzi, Earthquake-induced structural response output-only identification by two different operational modal analysis techniques. Earthquake Eng. & Struct. Dyn. **47**(1), 257–264 (2018)

45. M.L. Elliott, A.R. Knodt, D. Ireland, M.L. Morris, R. Poulton, S. Ramrakha, M.L. Sison, T.E. Moffitt, A. Caspi, A.R. Hariri, What is the test-retest reliability of common task-functional mri measures? New empirical evidence and a meta-analysis. Psychol. Sci. **31**(7), 792–806 (2020). pMID: 32489141. https://doi.org/10.1177/0956797620916786

46. S. Atasoy, I. Donnelly, J. Pearson, Human brain networks function in connectome-specific harmonic waves. Nat. Commun. **7**(1), 10340 (2016). https://doi.org/10.1038/ncomms10340

47. H.F. Iaccarino, A.C. Singer, A.J. Martorell, A. Rudenko, F. Gao, T.Z. Gillingham, H. Mathys, J. Seo, O. Kritskiy, F. Abdurrob, C. Adaikkan, R.G. Canter, R. Rueda, E.N. Brown, E.S. Boyden, L.-

H. Tsai, Gamma frequency entrainment attenuates amyloid load and modifies microglia. Nature **540**(7632), 230–235 (2016). https://doi.org/10.1038/nature20587

48. A.J. Martorell, A.L. Paulson, H.-J. Suk, F. Abdurrob, G.T. Drummond, W. Guan, J.Z. Young, D.N.-W. Kim, O. Kritskiy, S.J. Barker, V. Mangena, S.M. Prince, E.N. Brown, K. Chung, E.S. Boyden, A.C. Singer, L.-H. Tsai, Multi-sensory gamma stimulation ameliorates alzheimer's-associated pathology and improves cognition. Cell **177**(2), 256–271.e22 (2019). https://doi.org/10.1016/j.cell.2019.02.014

4.1   Research and Modeling Goals

The primary goal of this section is to develop a systematized modeling practice generating dynamic models using canonical state- space equations to describe the electrical activity in the brain from EEG data. We hypothesize that these models contain information that may be related to higher order cognitive states. The approach is especially focused on a modal decomposition of the state-space models, as modes effectively represent both the spatial and temporal information contained in EEG data. These generated eigenmodes may be related to human emotions that can be expressed in the existence of a finite number of discrete basis functions [1]. We hypothesize that an eigen-emotional state can be represented as an element of a Hilbert space, and these can be defined as a basis suitable for rigorous analysis. To prove the efficacy of our approach, we used EEG data from multiple publicly available EEG datasets in a cyber security application whose goal was to identify specific subjects based on their individual brain wave patterns. This classification model was developed from the output data using Output-only Modal Analysis (OMA) and Dynamic Mode Decomposition (DMD) in combination with a Neural Network. Further, our modeling practice is seen to confirm the existing knowledge regarding the existence of default mode networks [2].

This chapter follows with a detailed description of the proposed method in Sect. 4.2.1. Analysis of the modeling practice follows in Sect. 4.2.2. Then, the results of the classification problem are presented in Sect. 4.3, followed by a discussion of the overall outcomes in Sect. 4.4. We close with an overview of the remaining problems and future recommendations in Sect. 4.4.1.

Reprinted with permission from "System identification methods for dynamic models of brain activity", by T. Griffith, J. Hubbard, 2021. *Biomedical Signal Processing and Control* Vol. 68, pp. 102765, Copyright 2021 by Elsevier.

© The Author(s), under exclusive license to Springer Nature Switzerland AG 2023   65
T. D. Griffith et al., *A Modal Approach to the Space-Time Dynamics
of Cognitive Biomarkers*, Synthesis Lectures on Biomedical Engineering,
https://doi.org/10.1007/978-3-031-23529-0_4

4.2 Technical Approach to Cognitive Modeling

The following is a detailed description of the modeling procedure used herein for the extraction of dynamic modes as applied to EEG data. Two datasets were considered for analysis: the DEAP database and the EEGMMI database as described above in Section 3.2.2.

4.2.1 Adaptation of System Identification Algorithms for EEG Data

System identification algorithms have been used to generate dynamic models of complex systems across industries. Experimental modal analysis (EMA), in which the system of interest is artificially and precisely excited, can be used to determine the modal properties of the system [3]. EMA methods are not available for the study of EEG data, since the notion of a forced excitation to the brain is elusive at present. Many mechanical structures, such as large bridges, are also too complex to accept forced excitations [4]. Output-only modal analysis (OMA), was developed in the 1990s as a modal parameter estimation technique for structures under *imprecise operational excitations*. As OMA does not require any information regarding the input to the system of interest, we modified OMA for system identification of human EEG data. It is worth noting that OMA makes extensive use of an assumed broadband white noise input. Whether or not this is appropriate for EEG dynamics is a topic of some discussion but given the overall lack of distinct spikes in the power spectrum of any EEG signal, we feel it is appropriate. Further, the OMA modes generated perform well in the biosecurity classification task as seen in Sect. 4.3, which gives further credence to the approach. One of the main limitations of OMA is the inability to detect modes which are not naturally excited by the operating conditions of the unknown system. While this is an issue for mechanical structures, which may have unstable undetected modes, this is not an issue for EEG dynamics, which are not observed to diverge.

To contrast the OMA decomposition technique, a data-driven method is also considered. Since Dynamic Mode Decomposition (DMD) also directly yields an estimation of state-space dynamics and has been shown to perform well on ECoG data [5], it is selected for further analysis. The technical formulation and details of both OMA and DMD are well understood and can be found in [6, 7]. However, it is worth mentioning that both techniques yield models of the observed dynamics in state-space terms as

$$\begin{cases} \hat{x}(k+1) = A\hat{x}(k) \\ \hat{y}(k) = C\hat{x}(k) \end{cases}, \qquad (4.1)$$

where $x \in \mathbb{X}$ and $y \in \mathbb{Y}$. The matrix A estimates the internal dynamics, which we cannot measure, of the system based on the observations y. Note that, in this instance, the observations y are the measurable EEG data once it has been averaged to the common reference and bandpass. $\hat{y}$ is our best recreation of the true observable y. Due to uncertainty in the modeling procedure, and noise in the observations, we can define the error of our model as

$$e_y = \hat{y} - y. \tag{4.2}$$

A value of e_y in a neighborhood of zero indicates the matrix pair (A, C) is capturing the dynamics of the system well. The matrix C describes how the internal states combine to form the output. Crucially to the analysis of EEG dynamics, the matrix A may be decomposed as a series of modes, built around the eigenvectors of A. Each eigenvector v has an associated eigenvalue λ. We can rewrite any $x(k)$ as

$$x(k) = \sum_{i=1}^{n} \lambda_i^k v_i w_i^T \hat{x}(0), \tag{4.3}$$

which is a linear combination of modes. w_i are the left eigenvectors corresponding to λ_i. Equation (4.3) is referred to as the modal decomposition of the system. At every time step, the total dynamics may be represented as a weighted sum of modes. Each eigenvector v_i has a corresponding modal frequency λ_i. v_i is a vector describing the mode shape. Relative to the electrical activity in the brain, v_i elegantly illustrates how the different spatial regions of the brain impact one another at different temporal frequencies. Together, v_i and λ_i form an eigenmode. The eigenvectors v_i describe the spatial relationship at each frequency λ_i. It is our objective to identify high-fidelity modes which capture the spatial activity in the brain. Note that both the identification of A and the modal decomposition may be performed in near real time.

To adapt OMA for use on EEG data, spatial relationships must be defined for each of the sensors in the experimental setup. Koessler et al. have previously studied the statistical features of EEG electrode placement in Talairach space [8]. The average sensor locations were used here to define a three-dimensional model of the brain in space for each of the studied datasets. The model for the sensor locations in the DEAP dataset is shown in Fig. 4.1. Recorded EEG data is treated as an acceleration in the +z direction.

4.2.2 Analysis of EEG Eigenmodes

Having defined the spatial relationships of the sensors, OMA may be used on the datasets of interest. DMD does not require this explicit spatial representation. OMA and DMD decompose the raw EEG data into a series of modes, which we have defined as eigen-emotions in Eq. (4.3). Mode shapes may be visualized on the 3D model as shown in Fig. 4.2. Bruel & Kjaer's PULSE OMA Pro Type 8762 was used to develop the 3D model and compute the corresponding OMA modal decomposition.

It is often more convenient, however, to represent all the mode shapes at once on a complexity plot, as seen in Fig. 4.3. This figure shows the results of both modal decomposition

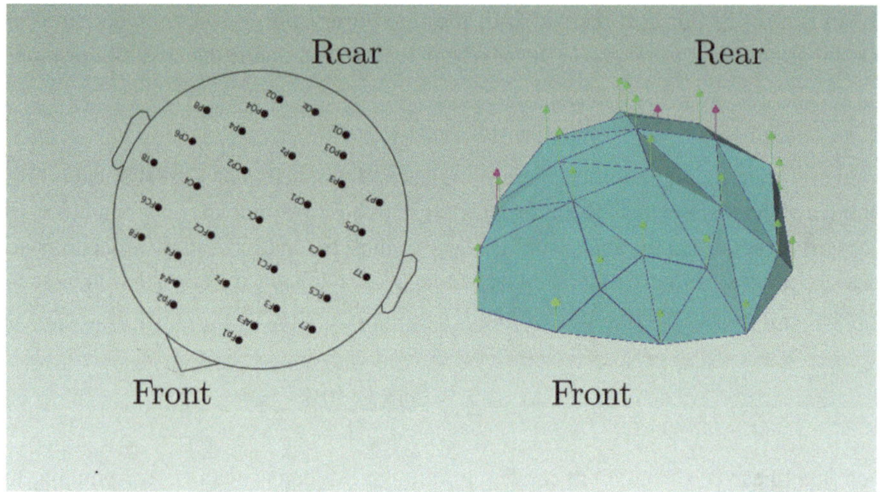

Fig. 4.1 Spatial representation of EEG sensor locations for DEAP dataset (Reproduced from [9])

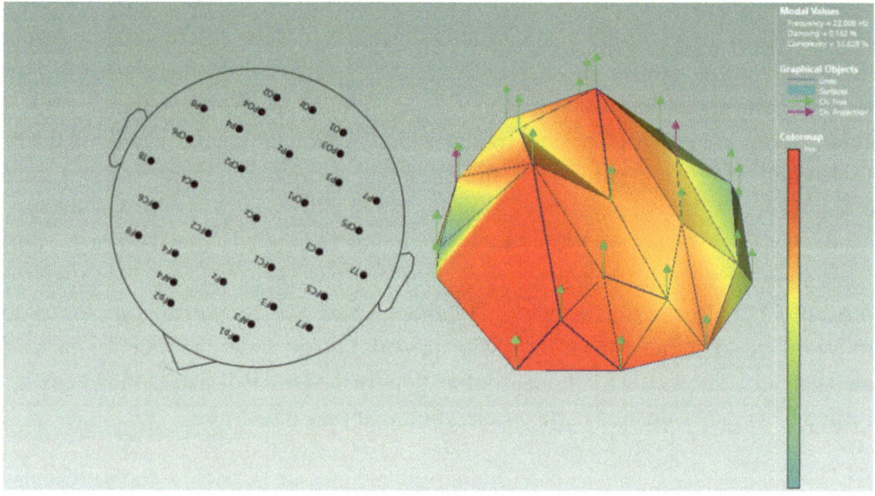

Fig. 4.2 A single in-phase eigen-emotion at peak amplitude from subject 17 (Reproduced from [9])

algorithms on a single trial in the DEAP database as an example. The high-fidelity modes are a linear decomposition of the observed measurement and form a basis for the emergent electrical activity. These complexity plots highlight the space-time nature of EEG dynamics.

The complexity plot assigns a magnitude, the length, and a phase, the angle, to each spatial location in the structure of interest. In Fig. 4.3, the 10 selected modes with the highest singular values of the A matrix are shown for simplicity, each in a different color. A mode associated with a high singular value is capturing more of the statistical variance in the data than the

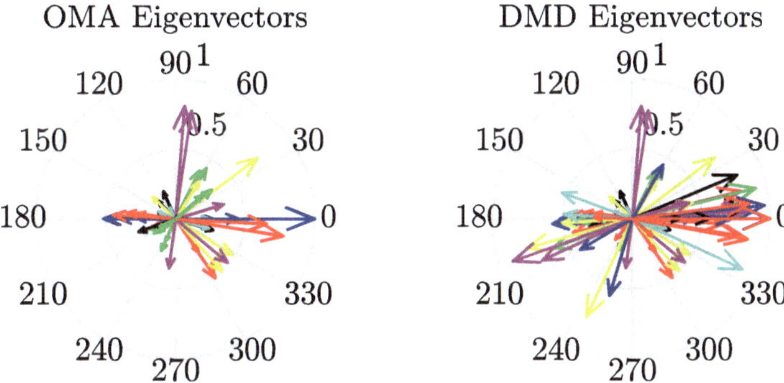

Fig. 4.3 Complexity plot example demonstrating both standing and traveling waves in EEG dynamics (Reproduced from [9])

modes with lower singular values. Arrows of common color share a temporal frequency but differ by spatial location. Notice that some modes, such as the blue mode in the OMA Eigenvector complexity plot, have zero phases. This corresponds to a standing wave in the oscillation of the mode. Each of the spatial locations reaches the maxima with total phase or antiphase at every point in time. Many mechanical structures, such as the Euler-Bernoulli beam, only show standing waves for example [10]. However, many of the modes generated from EEG data in the complexity plot here have a non-zero phase component. These modes display traveling waves and have complex valued eigenvectors. Although the eigenvectors in the system may be complex, they always come in conjugate pairs. This results in a real-valued state matrix which naturally generates real-valued data. While a given complex mode shares a temporal frequency, the spatial locations do not reach their peaks at the same time. The mode is out of phase, indicating a net energy flux from one location to another (i.e. a traveling wave across the EEG measurement aperture).

Notice further that the OMA algorithm extracts a few highly significant components, while DMD generates many components that have similar magnitude, but differ by phase. The modeling consequences of this difference in decomposition are seen in Sect. 4.3.1. One of the immediate consequences of this difference in system order involves the order reduction of the modal decomposition. Selecting the proper system is particularly important for OMA, which has known sensitivities to spurious noise modes [11]. Because the OMA algorithm involves a singular value decomposition, it is tempting to truncate the modal decomposition based on the singular values, as is common in DMD. However, because the OMA algorithm involves Hankel matrices, the number of singular values depends on the number of block rows included in the Hankel matrix. To determine the appropriate maximum system order, the singular values can be compared with the number of block rows as in Fig. 4.4.

Using the information in Fig. 4.4, the *maximum* system order can be selected. The singular values are either converging to zero or to a non-zero value as a function of the number of block rows. The maximum system order can be selected by choosing all non-zero singular values.

Fig. 4.4 Maximum order determination as a function of the number of block columns for DEAP data. Each line in a different color represents a singular value for a given row in the Hankel matrix as the number of block rows is increased (Reproduced from [9])

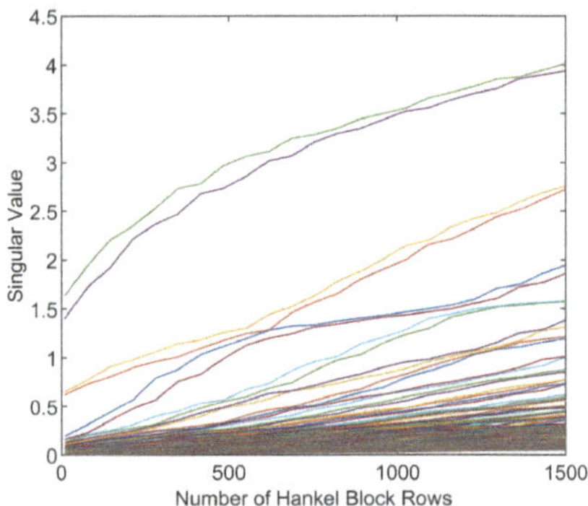

For the EEG data presented here, approximately 45 singular values are retained because they do not converge to zero. Unlike DMD, the number of singular values in the decomposition is not necessarily equal to the number of modes generated according to the standard OMA algorithm in [6]. Having selected a maximum system order, modes are generated. Then, we may choose to truncate the maximum order modal decomposition to an optimal number of modes. The r order state matrix may be truncated to nth order

$$A = \begin{matrix} n \\ r - n \end{matrix} \begin{bmatrix} A_{11} & A_{12} \\ A_{21} & A_{22} \end{bmatrix}. \tag{4.4}$$

The reduced order model is described by the smaller state matrix A_{11}. When truncating these OMA modes, the norm of the difference between the maximum order system, $H(z)$, and the reduced order model, $\hat{H}(z)$, is upper bounded by the neglected Hankel singular values through Enns's Conjecture [12]

$$||H(z) - \hat{H}(z)||_\infty \leq 2 \sum_{n+1}^{r} \sigma_k. \tag{4.5}$$

While the model could have been truncated at the singular value decomposition before modes had been computed for the maximum order system, the assurance of boundedness on the reconstruction error is mathematically lost. For the EEG data decomposed here, a 1% reconstruction error e_y was applied for system order truncation. On average, this results in approximately 25 distinct modes for the 63 s trials in the DEAP database.

Comparatively, model reduction for DMD algorithms is simpler. DMD also involves the use of singular value decomposition, but the matrix being decomposed is directly tied to the cardinality of the measured data. As a result, it is common practice to truncate modes which after 99% or 99.9% of the variance in the data has been captured, as in [7]. For the EEG data here, modes were truncated at the 99% threshold, generating just over 230 total modes on the 63 s trials as seen in Fig. 4.5. Note that because DMD involves a least squares

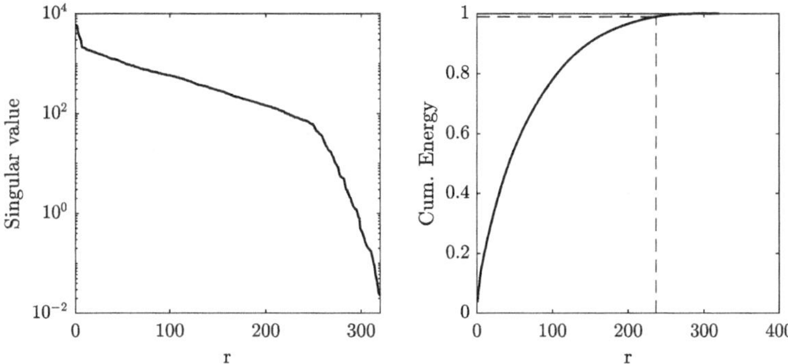

Fig. 4.5 Singular value vs. system order for subject 17 (Reproduced from [9])

regression for modal decomposition, complex modes come in pairs. As a result, only 115 distinct mode shapes are generated and used for modeling and analysis.

There is a non-trivial difference in the overall system order resulting from these decomposition techniques. DMD generates significantly more modes than OMA does. However, both methods are seen to effectively capture the information in the EEG waves in Sect. 4.3.1. From the described modeling procedure, we note the following observations.

4.2.3   The Existence of Stimuli Independent Common Modes

Across all subjects in the DEAP database, four common OMA modes were observed to be present. Regardless of the stimulus, we report the four modes in Table 4.1.

For the final column of the table, the Pearson correlation coefficient was calculated between the common modes for each of the 32 subjects individually and then averaged. This indicates that while the common modes have a near identical shape with a single subject, the mode shapes are not shared by all subjects. However, the modes share the similar temporal frequencies. All subjects have four common modes, near these frequencies, for all stimuli, but the emergent mode shape differs from subject to subject. We believe these common

Table 4.1 Common OMA brain modes of subjects in the DEAP database (Reproduced from [9])

	Frequency	Damping (%)	Complexity (%)	Shape correl.
Alpha mode 1	4.34±0.03	8.20±1.20	11.47±17.59	0.97±0.016
Beta mode 2	21.83±0.22	1.98±2.63	32.29±35.67	0.96±0.018
Gamma mode 3	40.39±0.26	11.87±7.49	12.42±16.88	0.99±0.010
Gamma mode 4	44.19±0.24	2.52±1.39	2.93±5.69	0.99±0.012

modes are connected to existing knowledge surrounding Default Mode Networks, which are active in specific frequency bands and vary among individuals [13, 14].

The remaining distinct modes do not correlate from stimuli to stimuli, even for the same subject. This further confirms existing knowledge that EEG signals vary over time scale and subject [15, 16]. It is natural to expect that these distinct modes are connected to emergent cognitive properties. We demonstrate the effectiveness of these distinct modes with a subject classification task.

4.3   Results of a Subject Classification Task

4.3.1   Experimental Validation and Verification Using a Neural Network Classifier

The generated modal decomposition was used to discriminate one subject from another using multiple datasets. In all demonstrations that follow, the only features provided for classification are the n eigenmodes (v_i, λ_i) which capture the emergent EEG dynamics in space and time. The eigenmodes determined from EEG data may be used to discriminate one subject from another using a neural network to achieve 100% accuracy with the OMA eigenmodes on a 5-fold cross validation. All multiclass classification accuracies reported here are on the cross validation sets are the total accuracy:

$$\text{Accuracy} = \frac{TP + TN}{TP + TN + FN + FP} \tag{4.6}$$

where TP is the number of true positives, TN is the number of true negatives, FN is the number of false negatives, and FP is the number of false positives for a given validation set. Here, we are demonstrating that the eigenmodes act as a brain wave subject identifier. Some regression technique is required to extract useful information for modeling from the modes. Machine learning techniques have recently shown good results on other highly variant data with spatial and temporal information encoded as an image [17–19]. This suggests that the same techniques may be applied to the extracted EEG eigenmodes. Note that the use of a neural network here serves only as a regression technique, where the data has first been transformed into a meaningful, canonical, mathematical model. The neural network does not discover the model but only analyzes it for trends that are difficult for a human to identify visually from the previously discussed complexity plots.

The relevant feature extraction and neural network algorithm are available in the form of MATLAB and Jupyter notebooks on Github (https://github.com/tdgriffith/SysID_ EEGdynamics). The generated eigenmodes can be transformed into an appropriate image. The important features in the eigenmodes can be organized into modal frequencies, spatial locations, and modal amplitudes as discussed earlier. These three degrees of freedom can then be converted into an image called here a heatmap. Figure 4.6 shows a reduced order

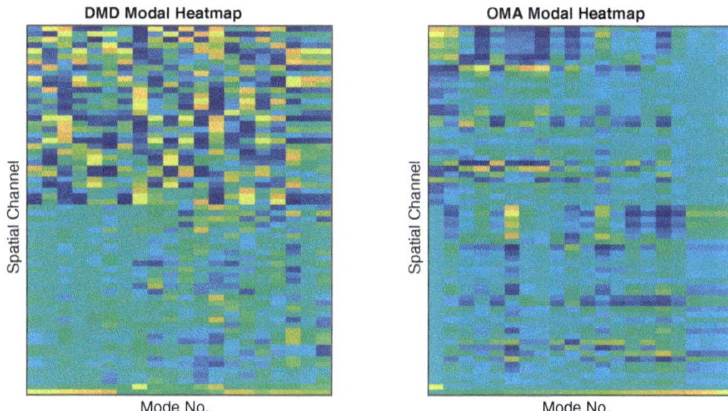

Fig. 4.6 Example heatmaps for subject classification (Reproduced from [9])

version of the eigenmode heatmaps as provided to the neural network for brain wave subject identification.

Having generated 1,280 heatmaps from the DEAP database (32 subjects, 40 trials), our modal decompositions were then encoded as images. The images may be input to a neural network for efficient classification. Since its inception in 2016, the ResNet architecture has served as a default method for computer vision machine learning [20, 21]. It is desirable to limit the complexity of the network to ensure that the system identification portion of the proposed method is doing most of the feature extraction [22]. For this reason, an 18-layer ResNet architecture was selected for the subject classification model. That is, the network includes 18 linear transforms with a nonlinearity in between each. 1024 heatmaps are in the training set, and the remaining 256 are reserved for validation. The learning rate was 1e−3, with a dropout rate of 10%. No data augmentation was performed, and the network was observed to train in fewer than 10 min total using the fastai library [23]. The last layer of the network was trained for 8 epochs, followed by the entire network for 5 epochs. The last layer was then trained for a final 5 epochs. *The 5-fold cross validation accuracy on the unseen data was 100% for the OMA modes and 98% for the DMD modes.* The relevant confusion matrices are included in Fig. 4.7.

4.3.2   An Extension to the EEG Motor Movement/Imagery Dataset

For each of the 109 subjects in the EEG Motor Movement/Imagery Dataset [24], 14 trials of various lengths were segmented into 20 s segments, to validate the approach on shorter time scales than the DEAP database. Since the segments are only 20 s long, some of the slow temporal dynamics in the EEG data may not be detectable by the OMA or DMD algorithms. The 83 modal heatmaps used were generated for each of the subjects, for a total of 9047 samples to train the subject identification network with. As in Sect. 4.3.1, a ResNet

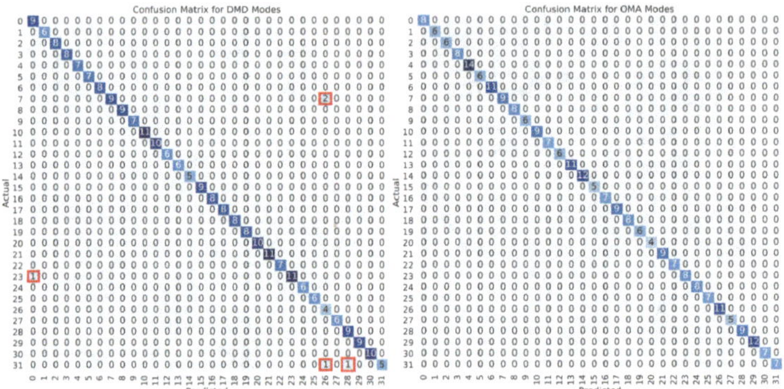

Fig. 4.7 Confusion matrices for subject identification on DEAP (Reproduced from [9])

architecture was used, but the number of layers was increased to 34 for better performance given the high number of subjects in the database. The learning rate was 1e−4, with a dropout rate of 10%. The last layer of the network was trained for 5 epochs, followed by the entire network for 10 epochs. Finally, the last layer was trained for 15 epochs. *The 5-fold cross validation accuracy was 96% for the DMD method, and 98% for the OMA method.*

4.3.3 Comparing OMA and DMD for Subject Identification

These classification results are promising. On the statistical nature of neural networks alone, validation results approaching 100% are very encouraging. On both databases, the OMA modes slightly outperformed the DMD modes. This is notable, especially when OMA generates approximately a third of the modes that DMD does. The mathematical structure and initial constraints of OMA appear to favor the information contained in EEG data. Furthermore, while the average mode shape, or average pixel value in the heatmaps, is similar for both algorithms, the OMA mode shapes are less variant than the DMD mode shapes. This suggests that while the data-driven DMD algorithm does a good job at capturing the high variance in the EEG data, the OMA algorithm captures fewer, highly descriptive modes. The OMA algorithm shows significantly longer wall times but offers a more efficient representation of the system for the classification task, eliminating some of the spurious noise modes. This may have implications for more advanced cognitive modeling outcomes, such as engagement or fatigue.

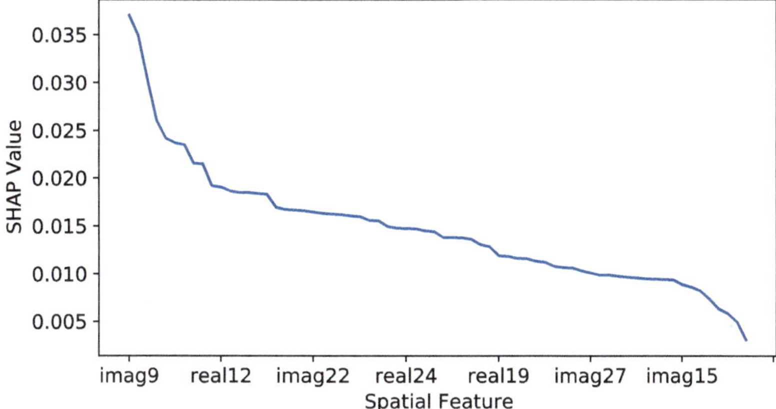

Fig. 4.8 SHAP values for DEAP subject identification (Reproduced from [9])

4.3.4 Optimal System Representations and Neural Network Interpretation

It is often valuable for applications to seek reductions in the system representation. For the purposes of our subject identification task, we are interested in performing the classification with the fewest number of physical sensors. Shapley Additive exPLanations use game theory to quantify how a single feature changes the output of a model through SHAP values [25, 26]. The SHAP value importance plot for the DEAP subject identification network is shown in Fig. 4.8.

With this notion of importance, we can select the most important channels for subject identification. As seen in Fig. 4.9, the minimum number of channels for $\geq 96\%$ accurate subject identification using the DEAP dataset is eight.

Fig. 4.9 Classification accuracy as a function of the number of spatial EEG channels (Reproduced from [9])

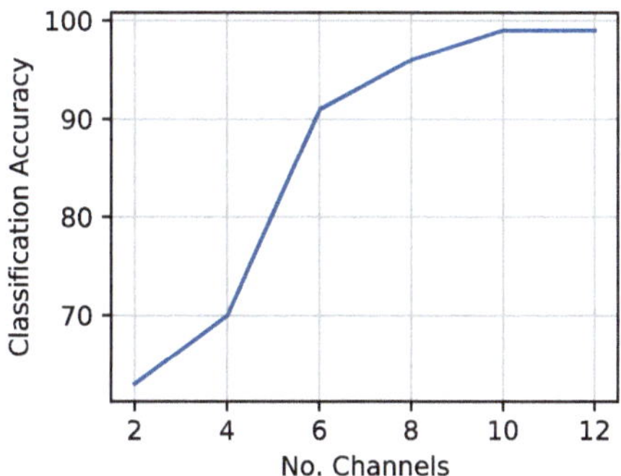

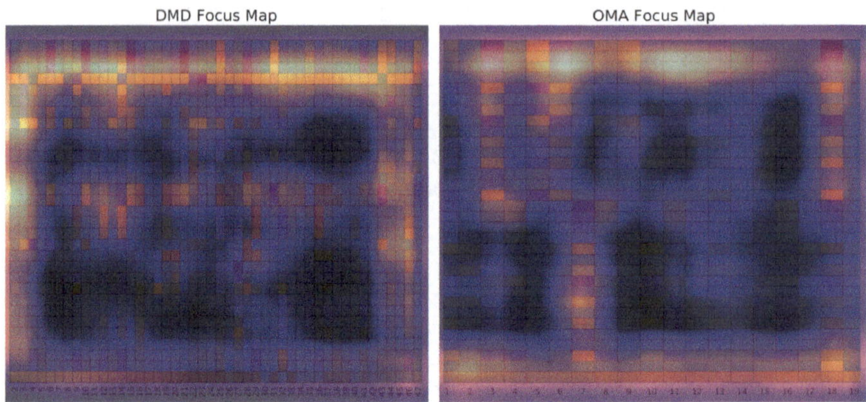

Fig. 4.10 Focus maps for OMA and DMD classification task (Reproduced from [9])

Since neural networks are a gradient method, they assign a weight to each pixel in the image. We can track these weights and overlay them on the original image. Hot spots indicate special attention was paid by the network to make a classification, while cold spots indicate the opposite. Interestingly, the OMA and DMD representations drive the networks to select similar spatial features for classification. As seen in Fig. 4.10, the subject identification network for the OMA modes observes a few specific columns in the heatmap. A single column corresponds to a spatial EEG channel location. At the same time, the DMD neural network takes a complete look at the edges of the image, which corresponds to the lowest and highest frequency modes. In both cases, the upper portion of the heatmap is critical, which corresponds to sensor locations toward the front of the skull. These results could be expected, given the modal decomposition as seen in Fig. 4.3, where the OMA algorithm was selecting a few very important mode shapes, while the DMD algorithm had many low norm modes to describe the overall dynamics.

Broadly, this gives initial credence to our system identification and modal approach to cognition and indicates that there is statistically significant information contained in these EEG eigenmodes.

4.3.4.1 Comparison to Existing Methods

It is worth comparing the performance of our eigenmode subject classification algorithm to existing results in the literature. Table 2 extends the analysis of [27] and compares subject identification accuracies on the DEAP database with our approach. We are encouraged by the success of [27], which achieved similar results using a more advanced deep learning architecture. That is, while our approach is to do more front-end modeling and train the network in relatively few (18) epochs, it is equally valid to employ a deeper network and

Table 4.2 Comparison of other subject identification results on the DEAP database (Reproduced from [9])

Reference	No. of electrodes	Accuracy (%)
Ours	32	100.00 ± 0.00
Ours	8	96.45 ± 0.14
[27]	32	99.90 ± 0.11
[27]	5	99.1 ± 0.34
[28]	32	97.97 ± 0.06
[29]	8	88 ± 4.0

[a] Table footnote (with superscript)

train for longer (185 epochs). This convergence of results suggests that both our eigenmodes and the network in [27] are uncovering an underlying structure in the EEG data which is relevant to subject identification.

4.4   Conclusions

A structured, ubiquitous approach for the identification of spatio-temporal dynamic modes of EEG brain activity was presented. Two system identification algorithms were adapted for use on EEG data, resulting in a modal representation of brain activity. The modal representation was shown to have complex eigenmodes representing traveling waves and net energy flow across the brain over time. Further, it was determined that all subjects have four unique brain modes that are present regardless of stimuli. We believe these common modes are connected to previously analyzed Default Mode Networks. Generated modal representations were converted to an image, from which a neural network could identify which subject the modes came from with nearly 100% accuracy. The OMA system identification technique was seen to slightly outperform DMD on the subject identification task. Further work to connect these eigen-emotions to emergent cognitive behaviors is required.

4.4.1   Recommendations

Future work extending this modal decomposition approach to the dynamics of human cognition, such as the estimation of fatigue or engagement, is expected. Special attention should be paid to the notion of first identifying a given subject, and then selecting an appropriate personalized model of that subject's cognition. Current markers of human outcomes heavily rely on self-reported data related to mood, affect, sleep patterns, and depressive symptoms, accumulated in an overall risk or performance score or rating. More recent approaches leverage signal-based indices of actigraphy, physiology, speech, and optical image recognition to provide more objective biomarkers of human outcomes. Despite the promising results, current literature depicts the following limitations: (1) self-reported and signal-based scores have been mostly examined in isolation, with few studies attempting to provide a holistic data driven approach to integrate them all [30]; (2) while the inherent inter-individual variability across people poses a significant challenge toward modeling human behavioral outcomes, previous models assume homogeneous patterns across individuals [31]. This technique offers a method of extracting relevant information from a variety of biomarkers that is mathematically rigorous. Since system identification techniques generally accept all forms of time series data, there is a further opportunity to extend this approach to multi-modal techniques using feature level collaborative filtering.

4.4.1.1 Estimating the Valence Arousal Scale from Eigenmodes

While we demonstrated in Sect. 4.3 that this linear, modal decomposition contains sufficient information to identify one subject from another, we were unable to successfully estimate the self-reported valence arousal rating provided by DEAP. Since we have rigorous mathematical justification for the modal decomposition here, it is natural to conclude that the nonlinear and time dependent effects of EEG brain waves need to be considered for emotional analysis.

More importantly, while the identification of linear modes has useful diagnostic information, as shown above, they do not describe unseen data well. That is, the modes extracted from a given time series do not generalize to other data from the same individual because of the nonstationary and nonlinear effects of EEG data.

In the following chapter, we will introduce the theoretical considerations for a robust adaptive state estimator which accounts for parametric uncertainty in the modes and estimates the unknown input to the brain wave plant. Crucially, this estimator updates the spatial pattern of the identified modes in real time to match the nonlinearities in the data.

References

1. H. Saarimäki, A. Gotsopoulos, I.P. Jääskeläinen, J. Lampinen, P. Vuilleumier, R. Hari, M. Sams, L. Nummenmaa, Discrete neural signatures of basic emotions. Cerebral Cortex **26**(6), 2563–2573 (2015). https://doi.org/10.1093/cercor/bhv086
2. L.E. Mak, L. Minuzzi, G. MacQueen, G. Hall, S.H. Kennedy, R. Milev, The default mode network in healthy individuals: a systematic review and meta-analysis. Brain Connect. **7**(1), 25–33 (2017). pMID: 27917679. https://doi.org/10.1089/brain.2016.0438
3. B.J. Schwarz, M.H. Richardson, Experimental modal analysis. CSI Reliab. Week **35**(1), 1–12 (1999)
4. R. Brincker, C. Ventura, *Introduction to Operational Modal Analysis* (Wiley, New York, 2015)
5. B.W. Brunton, L.A. Johnson, J.G. Ojemann, J.N. Kutz, Extracting spatial-temporal coherent patterns in large-scale neural recordings using dynamic mode decomposition. J. Neurosci. Methods **258**, 1–15 (2016). http://www.sciencedirect.com/science/article/pii/S0165027015003829

6. P. Van Overschee, B. De Moor, *Subspace Identification for Linear Systems: Theory-Implementation-Applications* (Springer Science & Business Media, Berlin, 2012)

7. J.N. Kutz, S.L. Brunton, B.W. Brunton, J.L. Proctor, *Dynamic Mode Decomposition* (Society for Industrial and Applied Mathematics, Philadelphia, PA, 2016). https://epubs.siam.org/doi/abs/10.1137/1.9781611974508

8. L. Koessler, L. Maillard, A. Benhadid, J. Vignal, J. Felblinger, H. Vespignani, M. Braun, Automated cortical projection of eeg sensors: anatomical correlation via the international 10–10 system. NeuroImage **46**(1), 64–72 (2009). http://www.sciencedirect.com/science/article/pii/S1053811909001475

9. T.D. Griffith, J.E. Hubbard, System identification methods for dynamic models of brain activity. Biomed. Signal Proc. Control **68**, 102765 (2021). https://www.sciencedirect.com/science/article/pii/S1746809421003621

10. J.N. Reddy, *Energy Principles and Variational Methods in Applied Mechanics* (Wiley, New York, 2017)

11. E. Reynders, System identification methods for (operational) modal analysis: review and comparison. Arch. Comput. Methods Eng. **19**(1), 51–124 (2012). https://doi.org/10.1007/s11831-012-9069-x

12. D.F. Enns, Model reduction with balanced realizations: an error bound and a frequency weighted generalization, in *The 23rd IEEE Conference on Decision and Control* (IEEE, 1984), pp. 127–132

13. C. Chang, G.H. Glover, Time–frequency dynamics of resting-state brain connectivity measured with fmri. NeuroImage **50**(1), 81–98 (2010). http://www.sciencedirect.com/science/article/pii/S1053811909012981

14. X. Leng, J. Xiang, Y. Yang, T. Yu, X. Qi, X. Zhang, S. Wu, Y. Wang, Frequency-specific changes in the default mode network in patients with cingulate gyrus epilepsy. Hum. Brain Map. **41**(9), 2447–2459 (2020). https://onlinelibrary.wiley.com/doi/abs/10.1002/hbm.24956

15. S. Makeig, C. Kothe, T. Mullen, N. Bigdely-Shamlo, Z. Zhang, K. Kreutz-Delgado, Evolving signal processing for brain–computer interfaces. Proc. IEEE **100**(Special Centennial Issue), 1567–1584 (2012)

16. L.I. Aftanas, A.A. Varlamov, S.V. Pavlov, V.P. Makhnev, N.V. Reva, Time-dependent cortical asymmetries induced by emotional arousal: eeg analysis of event-related synchronization and desynchronization in individually defined frequency bands. Int. J. Psychophysiol. **44**(1), 67–82 (2002). http://www.sciencedirect.com/science/article/pii/S0167876001001945

17. W. Ke, Y. Xing, G. Di Caterina, L. Petropoulakis, J. Soraghan, Intersected emg heatmaps and deep learning based gesture recognition, in *Proceedings of the 2020 12th International Conference on Machine Learning and Computing*, ICMLC 2020 (Association for Computing Machinery, New York, NY, USA, 2020), pp. 73–78. https://doi.org/10.1145/3383972.3383982

18. V.K. Venugopal, K. Vaidhya, M. Murugavel, A. Chunduru, V. Mahajan, S. Vaidya, D. Mahra, A. Rangasai, H. Mahajan, Unboxing ai - radiological insights into a deep neural network for lung nodule characterization. Acad. Radiol. **27**(1), 88–95 (2020). Special Issue: Artificial Intelligence. http://www.sciencedirect.com/science/article/pii/S1076633219304489

19. K. Tangsali, V.R. Krishnamurthy, Z. Hasnain, Generalizability of convolutional encoder-decoder networks for aerodynamic flow-field prediction across geometric and physical-fluidic variations. J. Mech. Design **143**(5), 051704 (2020). https://doi.org/10.1115/1.4048221

20. K. He, X. Zhang, S. Ren, J. Sun, Deep residual learning for image recognition, in *Proceedings of the IEEE Conference on Computer Vision and Pattern Recognition (CVPR)* (2016)

21. C. Zhang, P. Benz, D.M. Argaw, S. Lee, J. Kim, F. Rameau, J.-C. Bazin, I.S. Kweon, Resnet or densenet? introducing dense shortcuts to resnet, in *Proceedings of the IEEE/CVF Winter Conference on Applications of Computer Vision (WACV)* (2021), pp. 3550–3559

22. S. Koning, C. Greeven, E.O. Postma, Reducing artificial neural network complexity: a case study on exoplanet detection (2019). arXiv:abs/1902.10385
23. J. Howard, S. Gugger, Fastai: a layered api for deep learning. Information **11**(2), 108 (2020). http://dx.doi.org/10.3390/info11020108
24. G. Schalk, D.J. McFarland, T. Hinterberger, N. Birbaumer, J.R. Wolpaw, Bci 2000: a general-purpose brain-computer interface (bci) system. IEEE Trans. Biomed. Eng. **51**(6), 1034–1043 (2004)
25. M.T. Ribeiro, S. Singh, C. Guestrin, "why should i trust you?": Explaining the predictions of any classifier, in *Proceedings of the 22nd ACM SIGKDD International Conference on Knowledge Discovery and Data Mining*, KDD '16 (Association for Computing Machinery, New York, NY, USA, 2016), pp. 1135–1144. https://doi-org.srv-proxy1.library.tamu.edu/10.1145/2939672.2939778
26. Q. Dickinson, J.G. Meyer, Positional shap for interpretation of deep learning models trained from biological sequences. *bioRxiv* (2021) https://www.biorxiv.org/content/early/2021/03/05/2021.03.04.433939
27. T. Wilaiprasitporn, A. Ditthapron, K. Matchaparn, T. Tongbuasirilai, N. Banluesombatkul, E. Chuangsuwanich, Affective eeg-based person identification using the deep learning approach. IEEE Trans. Cognit. Dev. Syst. **12**(3), 486–496 (2020)
28. M. DelPozo-Banos, C.M. Travieso, J.B. Alonso, A. John, Evidence of a task-independent neural signature in the spectral shape of the electroencephalogram. Int. J. Neural Syst. **28**(01), 1750035 (2018). pMID: 28835183. https://doi.org/10.1142/S0129065717500356
29. Y. Li, Y. Zhao, T. Tan, N. Liu, Y. Fang, Personal identification based on content-independent eeg signal analysis, in *Biometric Recognition*. ed. by J. Zhou, Y. Wang, Z. Sun, Y. Xu, L. Shen, J. Feng, S. Shan, Y. Qiao, Z. Guo, S. Yu (Springer International Publishing, Cham, 2017), pp.537–544
30. E. Debie, R.F. Rojas, J. Fidock, M. Barlow, K. Kasmarik, S. Anavatti, M. Garratt, H.A. Abbass, Multimodal fusion for objective assessment of cognitive workload: a review. IEEE Trans. Cybern. 1–14 (2019)
31. H.P.A. Van Dongen, G. Maislin, D.F. Dinges, Dealing with inter-individual differences in the temporal dynamics of fatigue and performance: importance and techniques. Aviation Space Environ. Med. **75**(3), A147–A154 (2004). https://www.ingentaconnect.com/content/asma/asem/2004/00000075/a00103s1/art00025

Adaptive Unknown Input Estimators

5

In Chap. 4, we demonstrated a technical approach to modeling the linear behavior of EEG brain waves with a modal decomposition generated by system identification techniques. Analysis of these modes was presented, but we were unable to estimate the self-report valence arousal ratings in the DEAP database successfully. In order to improve the fidelity of these models to achieve better accuracy on the DEAP database, we developed an adaptive unknown input observer *to treat parametric uncertainty in the modes, deal with nonlinear effects in the data*, and simultaneously form a real-time estimate of the EEG brain wave system input subject to a series of constraints. This estimate of the input grasps the exogenous information to the system which may not be generated by the modes and improves our estimator performance. This chapter presents the theoretical details of our adaptive unknown input estimator. Since it is heavily focused on the stability and performance analysis of the estimator architecture, it is somewhat removed from the analysis of spatio-temporal brain wave dynamics until Chap. 6. This estimator is highly nonlinear, so ensuring its stability is crucial. We will begin with the most basic form of the adaptive unknown input observer, before proceeding to introduce subsequent developments which improved the fidelity of the estimation process.

5.1 Introduction

Estimating the internal state of a linear time-invariant dynamical system has generated a tremendous amount of knowledge in linear systems and control [1, 2]. Conventional state estimators assume access to all system inputs for all time. In some dynamical processes,

---

*Reprinted with permission from "An Adaptive Control Framework for Unknown Input Estimation," by T. Griffith, M. Balas, 2021. *Proceedings of the ASME 2021 International Mechanical Engineering Congress and Exposition* Copyright 2021 by ASME.

however, the system inputs may not be known because it may be difficult to place sensors [3] or recreate the operating conditions which provoke a given input [4, 5]. As a result, much research effort has considered the estimation of an unknown input simultaneously with the estimation of the system state [6–9].

Approaches to this problem have typically involved using some information already available in the estimator to determine the unknown input [10] or including an internal model of the unknown input waveform [11]. By including an internal model of the waveform, the estimator gives reliable online estimates of the internal dynamics and the individual waveforms of the unknown input, which can improve controller performance as discussed in our previous work [12]. Others have used waveform generators extensively for disturbance accommodation and control [13–15]. The approach may be extended to unknown input estimation.

More generally, these unknown input estimators are often developed for systems with significant uncertainty or expected faults, requiring robust estimation schemes. In [16], a nonlinear unknown input observer is used to decouple the input from the estimated state in a robust manner. Several extensions to this idea have generalized it and achieved performance improvements [17, 18]. These robust approaches generally focus on fault detection and isolation, rather than explicit state estimation for control [19].

Adaptive control laws can also account for some uncertainty in the dynamical model, expanding the operating neighborhood of the linear time-invariant system. A robust and adaptive sliding mode controller is implemented in [20] to handle nonlinearity and uncertainty. Further, it was shown in [21] that a bank of linear time-invariant systems could be used for the detection of a variety of faults. In [3], an adaptive estimator is used to capture the time-varying dynamics of an internal combustion engine. More recently, in [10], a general adaptive control architecture was presented for the estimation of an unknown input with convergence guarantees for a wide class of systems. In each of these cases, the estimator must treat the inherent uncertainty of the model. This uncertainty in the dynamics is especially prevalent in models generated with system identification techniques and parameter estimation, where unique representations of the dynamics may be inaccessible [22].

In such cases, a control architecture that not only converges the state and input errors, but also recovers the physical structure of the plant dynamics is desirable. This chapter presents a novel architecture for the adaptive estimation of unknown inputs. An input generator, in the form of an ordinary differential equation (ODE), is implemented which models the unknown input as a linear combination of basis functions. As introduced in [12], this model estimates the coefficients of each basis function which may contain important diagnostic information. A time-varying control law is implemented to recover the physical structure of the model. The main contribution of the chapter is the control architecture shown in Fig. 5.1 along with a proof of error convergence under assumptions that are applicable to a wide class of dynamical systems.

We begin with a description of the class of systems considered by our adaptive unknown input estimator. We assume the system is almost strictly dissipative (ASD) or strictly dissi-

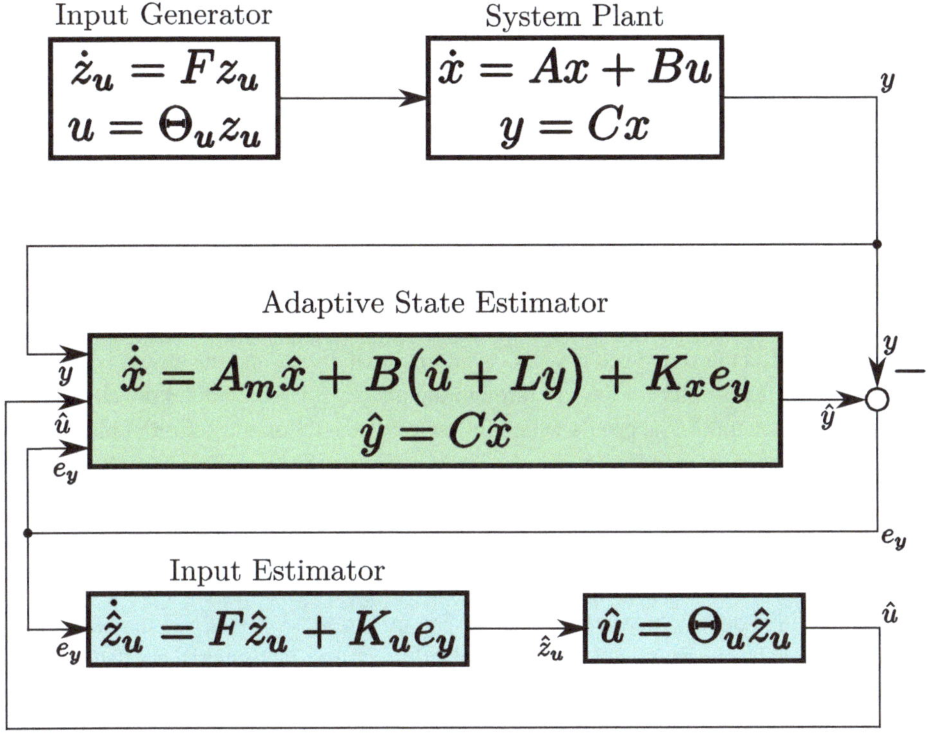

Fig. 5.1 The adaptive unknown input estimator. (Reproduced from [23])

pative (SD), and has a bounded output. For the modeling of brain wave dynamics, this is not a particularly restrictive condition since the brain wave dynamics are stable without external control input. The ASD property is crucial to use Lyapunov stability analysis on adaptive nonlinear estimators. The estimator uses the output relative error to recover the physical structure of the internal dynamics and unknown input. We claim that the architecture in Fig. 5.1 reliably estimates both the internal plant states and unknown input states in real time. Later, we will present the proof that substantiates this claim.

This overview of the system dynamics is followed by a discussion of the input generator approach to input estimation. While the input generator offers advantages for estimator synthesis and convergence results, the performance of the estimator is sensitive to the formulation of the input generator. Waveforms not considered are simply unaccounted for, which creates an opportunity for undetectable inputs. Then, the main result of the chapter is presented, which includes the formulation of our adaptive unknown input estimator and convergence proofs.

5.2   Unknown Input Dynamics

The proposed architecture seeks to build on the existing body of knowledge surrounding unknown input estimation by using adaptive estimators to account for uncertain internal dynamics, especially uncertain dynamics arising from the use of system identification techniques. Consider the class of linear dynamical plants

$$\begin{cases} \dot{x}(t) = Ax(t) + Bu(t) \\ y(t) = Cx(t), \end{cases} \tag{5.1}$$

where $x(t) \in \mathbb{X}$ is the state vector, $u \in \mathbb{U}$ is an unknown input, and the output is $y(t) \in \mathbb{Y}$. The unknown inputs enter the system through the matrix B which defines the channels the unknown inputs act on. The proofs which follow rely heavily on the notion of almost strict dissipativity. The system in Eq. (5.1) is ASD if it meets the Kalman-Yakubovich conditions

$$A_c^* P + P A_c = -Q \tag{5.2}$$
$$P B = C^*$$

for some positive definite matrices P and Q. The matrix $A_c = A + BGC$ may stabilize A with a stable gain matrix G and is equivalent to the closed loop matrix for the system in Eq. (5.1) under the feedback law $u = Gy$. The ASD property is equivalent to the open loop system having a positive definite high-frequency gain and stable minimum phase transmission zeros [24]. This property is crucial in adaptive systems when using Lyapunov-based analysis. It is reasonable to expect that the ASD property will hold for the systems of interest. If a system has an unknown input, and we are still able to collect output measurements y, we might expect the system to be ASD since it does not diverge in the presence of a potentially stochastic unknown input. This would occur if the plant itself was strictly dissipative, or if a known input was employed alongside the unknown input to stabilize the system.

The true input, about which we have only limited knowledge, is given by the vector function $u(t)$. Our modeling approach for $u(t)$ is to approximate its space $\mathbb{U}$ with an nth order linear combination as

$$\hat{u}(t) = \sum_{i=1}^{N} c_i f_i(t). \tag{5.3}$$

Equation (5.3) should only be considered a function space approximation of the unknown input. The individual functions f_i, which we will refer to as basis functions, are determined beforehand based on some knowledge of the system and unknown input. The weighting coefficients c_i are completely unknown and may be discontinuous at different times while the estimator is active. By including some knowledge about the unknown input, we gain the required structure to develop convergence proofs as in [4].

We further consider the model presented in Eq. (5.3), where $\{f_1, f_2, \ldots, f_N\}$ are linearly independent known functions in $\mathbb{U}$ and the coefficients c_i are scalars. The coefficients c_i are chosen to minimize the input error $E \equiv ||u - \hat{u}||^2$ in the Hilbert space $\mathbb{U}$ of inputs to the dynamic system in Eq. (5.1). These dynamics can be rewritten in this view as

$$\begin{cases} \dot{x} = Ax + B\hat{u} + \epsilon_u \\ y = Cx \end{cases} \tag{5.4}$$

with the input estimation error $\epsilon_u \equiv u - \hat{u}$. If we had complete knowledge of $u(t)$, we could use well studied methods of approximation theory to obtain the best approximation of the N-dimensional subspace $S_N \equiv \text{span}\{f_1, f_2, \ldots, f_N\}$ to minimize E [25–27]. The most widely known solutions to this classic problem involve least squares approaches primarily. Of course, we do not know $u(t)$, so we are forced to initially make a general approximation of the waveform or signal shape as in Eq. (5.3). It is our objective to reliably estimate c_i online with our input state estimator.

5.3   Input Generators as a Model of the Unknown Input

To estimate an unknown input using the known waveform approach, an internal model of the waveform must be included in the estimator. Therefore, Eq. (5.3) must be converted into a set of first-order differential equations which may be cast in state space [11]. For performance and stability reasons, it is desirable for the state-space input generator to be a linear time-invariant system. These input generators take the form

$$\begin{cases} \dot{z}_u(t) = F_u z_u(t) \\ \hat{u}(t) = \Theta_u z_u(t). \end{cases} \tag{5.5}$$

Both F_u and Θ_u are completely known matrices. We will refer to z_u as the state of $\hat{u}(t)$, which are the individual components of the estimated unknown input.

Note that not all bases $\{f_1, f_2, \ldots, f_N\} \in \mathbb{U}$ can be generated by the input generator as formed in Eq. (5.5). For example, wavelets are not causal and may not be generated with a linear time-invariant model. Therefore, it is worth discussing our process for generating bases for unknown input estimations. The set of bases available to us consists of those that may be generated by an N-order differential equation

$$L\hat{u} = a_0\hat{u} + a_1\hat{u}^{(1)} + \cdots + a_{N-1}\hat{u}^{(N-1)} + \hat{u}^{(N)} = 0, \tag{5.6}$$

where a_k may be real or complex scalars and $\hat{u}^{(k)} = \frac{d^k\hat{u}}{dt^k}$. It is well known from [28] that if $\{f_1, f_2, \ldots, f_N\}$ are chosen so they belong to the null space of L, $N(L) = \{x \in \mathbb{U} | Lx = 0\}$, and satisfy the Wronskian

$$W(t) = \begin{bmatrix} f_1 & f_2 & \cdots & f_N \\ f_1^N & f_2^N & \cdots & f_N^N \\ \vdots & \vdots & \ddots & \vdots \\ f_1^{N-1} & f_2^{N-1} & \cdots & f_N^{N-1} \end{bmatrix} \begin{bmatrix} c_1 \\ c_2 \\ \vdots \\ c_N \end{bmatrix} \tag{5.7}$$

as being nonsingular at $t = 0$, then they form a basis for $N(L)$ and all input approximations can be uniquely represented in $S_n \equiv N(L)$ with Eq. (5.3) by Abel's Theorem [29]. As an example, consider

$$\begin{cases} \dot{z}_u(t) = \begin{bmatrix} 0 & 1 \\ -\omega^2 & 0 \end{bmatrix} z_u(t) \\ \hat{u}(t) = \begin{bmatrix} 1 & 1 \end{bmatrix} z_u(t) \end{cases} \tag{5.8}$$

which will generate an input of the form $u(t) = \sin(\omega t) + \cos(\omega t)$. Input generators of this form are well studied, and details can be found in [11, 30, 31]. This implies that our adaptive state estimator and unknown input estimator can recover the approximate input online in real time along with the dynamic system model states, provided the input can be written as a linear time-invariant state-space system. Therefore, the chosen basis $\{f_1, f_2, \ldots, f_N\}$ from Eq. (5.6) is a critical element for obtaining a high-fidelity approximation $\hat{u}(t)$ of the true unknown input $u(t)$. The choice of the basis to do this is governed by how much of the information in the true input can be captured by the N terms of the estimated approximate input. The Shannon Entropy for continuous channels from Information Theory can aid in the choice of the input generator basis set $\{f_i\}_{i=1}^{N}$ that maximizes the information content captured by $\hat{u}(t)$ when N is small [32, 33].

Practically, f_i should be chosen to match observations of $u(t)$. Some information about the expected behavior of $u(t)$ may be available in offline observations of the system's environment. For example, as we have alluded to in Eq. (5.8), an unknown input with observed periodic tendencies should be modeled with a set of Fourier components. Crucially, f_i should always include a step function generator to model the mean of the unknown input. In cases where knowledge of $u(t)$ is completely unknown, we recommend the heuristic in [34], which is a polynomial spline model of the form

$$u(t) = c_1 + c_2 t + c_3 t^2 + \cdots + c_n t^{n-1}. \tag{5.9}$$

A polynomial order of $n = 4$ has historically been sufficient for estimation and controller design. Further, f_i is not restricted to only elements of a single basis. For example, a fourth order polynomial spline and Fourier components could be simultaneously generated by F_u.

5.4   Main Result: Adaptive Control Architecture for Unknown Input Estimation

We begin by considering the same plant in Eq. (5.1) and input generator in Eq. (5.5). New to the formulation we introduce a distinction between the state matrix for the plant, which represents the true dynamics of the system as A, and the state matrix for the model, which represents our best recreation of A from system identification or parameter estimation as A_m.

$$\text{TRUE PLANT:} \begin{cases} \dot{x} = Ax + Bu \\ y = Cx \end{cases} \tag{5.10}$$

$$\text{UNCERTAIN MODEL:} \begin{cases} \dot{x}_m = A_m x + Bu \\ y_m = C x_m. \end{cases} \tag{5.11}$$

Again, and crucially $A \neq A_m$. We may only use the observable information in the system to correct A_m. It is common practice in adaptive control to include an additional feedback term Ly to the state estimator, where the feedback gain matrix L is time varying.

$$\text{STATE EST.:} \begin{cases} \dot{\hat{x}} = A_m \hat{x} + B\left(\Theta_u \hat{z}_u + Ly\right) + K_x e_y \\ \hat{y} = C\hat{x} \end{cases} \tag{5.12}$$

$$\text{INPUT EST.:} \begin{cases} \dot{\hat{z}}_u = F\hat{z}_u + K_u e_y \\ \hat{u}_D = \Theta_u \hat{z}_u. \end{cases} \tag{5.13}$$

This feedback term can be thought of in two ways. First, we may consider the entire feedback term as the unknown input $\hat{u} = \Theta_u z_u + Ly$. In this view, the uncertainty in the model is compensated for by simply adjusting the unknown input so that the model output is the same as the output of the plant. Alternatively, we may return to the formulation of the unknown input as simply $\hat{u} = \Theta_u z_u$. Then, we rewrite the adaptive feedback as $Ly = LCx$, which corrects A_m if we assume there is some optimal feedback matrix L_* that results in $A \equiv A_m + BL_*C$. When L converges to L_*, the physical structure of the plant is recovered. While these views are mathematically equivalent, we prefer to conceptualize this scheme as converging $L \to L_*$ to correct the uncertainty in A_m and so consider the estimated input $\hat{u} = \Theta_u z_u$.

Since it is time variant, and there is a target value, the feedback gain matrix L has its own error matrix ΔL

$$\Delta L = L - L_*, \tag{5.14}$$

where the gain L is adaptive

$$\Delta \dot{L} = \dot{L} = -e_y y^* \gamma_e; \quad \gamma_e > 0. \tag{5.15}$$

In addition to correcting the uncertainty in the model, the adaptive gain introduces a non-linearity to the control scheme. Almost all systems exhibit some nonlinear effects. The adaptive feedback here serves to expand the neighborhood for which A_m is valid. γ_e is a tunable parameter that scales the change in L. A greater value of γ_e expands the operating neighborhood of the model but may cause the system to diverge if it is too responsive to estimator error.

5.4.1   Composite Error Dynamics

Given this estimator scheme, the error dynamics for the composite system $\bar{x} = \begin{bmatrix} x \\ z_u \end{bmatrix}$ are

$$\dot{e} = (\bar{A} + \bar{K}\bar{C})e + \bar{B}\Delta Ly \tag{5.16}$$

$$\dot{e} = (\bar{A} + \bar{K}\bar{C})e + \bar{B}w \tag{5.17}$$

$$\begin{bmatrix} \dot{e}_x \\ \dot{e}_z \end{bmatrix} = \left(\begin{bmatrix} A_m & B\Theta_u \\ 0 & F \end{bmatrix} + \begin{bmatrix} K_x \\ K_u \end{bmatrix} \begin{bmatrix} C & 0 \end{bmatrix} \right) \begin{bmatrix} e_x \\ e_z \end{bmatrix} + \begin{bmatrix} B \\ 0 \end{bmatrix} w \tag{5.18}$$

$$= \underbrace{\begin{bmatrix} A_m + K_x C & B\Theta_u \\ K_u C & F \end{bmatrix}}_{\bar{A}_c} \begin{bmatrix} e_x \\ e_z \end{bmatrix} + \begin{bmatrix} B \\ 0 \end{bmatrix} w \tag{5.19}$$

$$\text{with, } e_y = \begin{bmatrix} C & 0 \end{bmatrix} \begin{bmatrix} e_x \\ e_z \end{bmatrix}. \tag{5.20}$$

Here $w \equiv \Delta Ly$. Notice that the gain matrix $\bar{K}$ is not adaptive and must be placed with standard methods such as linear quadratic regulators to stabilize the eigenvalues of $\bar{A}_c$. With successful convergence results, the scheme in Eqs. (5.12) and (5.13) will provide reliable real-time estimates of x, z_u, and u, all while correcting the uncertainty in A_m. This novel scheme has wide application to systems with uncertainty in the modeling process and control synthesization.

5.4.2   Proof of Composite Error Convergence

It is desirable for L to converge to L_* while also estimating the unknown input, so the pair $(\bar{A}, \bar{C})$ must be observable. Even though the input estimator alone is often unobservable, the matrix pair $(\bar{A}, \bar{C})$ can be completely observable, which implies that estimates of both the state and input can be implemented in real time. Consider $V(e, \Delta L)$, the composite Lyapunov function for the state error e and the gain error ΔL

$$V(e, \Delta L) = \frac{1}{2}e^* \bar{P} e + \frac{1}{2}tr(\Delta L \gamma_e^{-1} \Delta L^*), \tag{5.21}$$

where $\bar{P}$ and γ_e are symmetric positive definite matrices, and $tr(\bullet)$ is the matrix trace operator. The time derivative of $V(e, \Delta L)$ is

$$\dot{V}(e, \Delta L) = e^* \bar{P} \dot{e} + tr(\Delta \dot{L} \gamma_e^{-1} \Delta L^*) \tag{5.22}$$
$$= e^* \bar{P}(\bar{A}_c e + B \Delta L y) + tr((-e_y y^* \gamma_e)\gamma_e^{-1} \Delta L^*) \tag{5.23}$$
$$= e^* \bar{P} \bar{A}_c e + e^* \bar{P} \bar{B} \Delta L y + tr(-e_y y^* \Delta L^*) \tag{5.24}$$
$$= \frac{1}{2}e^*(\bar{P}\bar{A}_c + \bar{A}_c^* \bar{P})e + e^* \bar{P} \bar{B} w - (e_y, w). \tag{5.25}$$

Using the assumed ASD property of the plant, we know that

$$\bar{P}\bar{A}_c + \bar{A}_c \bar{P} = -\bar{Q} \tag{5.26}$$
$$\bar{P}\bar{B} = \bar{C}^* \tag{5.27}$$

and Eq. (5.25) can be reduced to

$$\dot{V}(e, \Delta L) = -\frac{1}{2}e^* Q e + \underbrace{e^* \bar{C}^* w}_{(} e_y, w) - (e_y, w) \tag{5.28}$$

$$= -\frac{1}{2}e^* Q e \tag{5.29}$$

$$\leq -\frac{1}{2}\lambda_{\min}(Q)||e||^2. \tag{5.30}$$

Equation (5.30) bounds the trajectory $e(t)$ and ΔL. Now consider the function

$$W(e) \equiv \frac{1}{2}\lambda_{\min}(Q)||e||^2 \tag{5.31}$$

and take the time derivative yielding

$$|\dot{W}(e)| = \lambda_{\min}(Q)|e^* \dot{e}|. \tag{5.32}$$

Substitute $\dot{e}$ from Eq. (5.16) to see that

$$|\dot{W}(e)| = \lambda_{\min}(Q)\left|e^*\left[(\bar{A} + \bar{K}\bar{C})e + \bar{B}\Delta L y\right]\right|. \tag{5.33}$$

While Eq. (5.30) bounds e and ΔL, it does not bound y. We employ the assumption that as a property of the plant, the output state y is bounded. Given this, the derivative of Eq. (5.31) is also bounded, and $W(e)$ is uniformly continuous by the mean value theorem. Finally, by Barbalat's lemma, $e(t) \to 0$ as $t \to \infty$.

This result guarantees the convergence of the entire vector $e = \begin{bmatrix} e_x & e_z \end{bmatrix}^T$. The error in both the state and the unknown input state converges to zeros as $t \to \infty$. Notice that while the proof guarantees convergence in e_z, not just e_u, it does not guarantee a convergence rate for e. Further, although L is bounded, there is no guarantee of convergence to L_*.

5.5   Illustrative Examples

A simple example is included for numerical demonstration of the outlined architecture. Simulation files and results are available at the URL address https://github.com/tdgriffith/IMECE_UIO. We treat a system as in Eq. (5.11), with

$$A_m = \begin{bmatrix} -7 & 2 & 4 \\ -2 & -1 & 2 \\ -2 & 2 & 1 \end{bmatrix}, \quad B = \begin{bmatrix} 0 \\ 0.7 \\ 2 \end{bmatrix}, \quad C = \begin{bmatrix} 0.5 & 0 & 1 \end{bmatrix}. \tag{5.34}$$

Recall that this matrix triple (A_m, B, C) has some uncertainty associated with it. For the purposes of demonstration, we assign

$$L_* = \begin{bmatrix} -5 \end{bmatrix}. \tag{5.35}$$

so that the plant has the dynamics

$$A = \begin{bmatrix} -7 & 2 & 4 \\ -3.75 & -1 & -1.5 \\ -7 & 2 & -11 \end{bmatrix}, \quad B = \begin{bmatrix} 0 \\ 0.7 \\ 2 \end{bmatrix}, \quad C = \begin{bmatrix} 0.5 & 0 & 1 \end{bmatrix}. \tag{5.36}$$

The dynamics in Eq. (5.34) have different eigenvalues and eigenvectors from the true dynamics in Eq. (5.36) as evidenced by Fig. 5.2.

Fig. 5.2 Impulse response of identified system compared to the true plant. (Reproduced from [23])

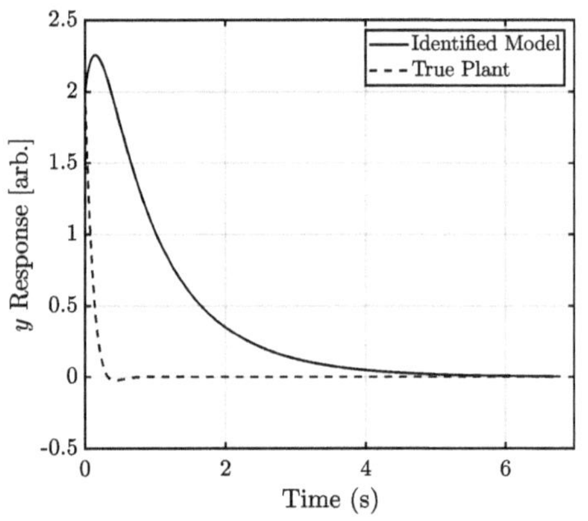

Both systems have two stable transmission zeros. Then, by [24], both systems meet the Kalman-Yakubovich conditions and are ASD. Given a bounded output y, our scheme in Eqs. (5.12) and (5.13) is guaranteed to converge with an appropriate input generator. Suppose the unknown input takes the following waveform:

$$u_1(t) = c_1 + c_2 \sin(2t) + c_3 \cos(4t), \tag{5.37}$$

where the constants c_1, c_2, and c_3 are unknown to the input estimator. For the purposes of demonstration, we set them to 3, 2, and 4, respectively. An input generator for the entire unknown input takes the form

$$\begin{cases} \dot{\hat{z}}_u = F_u z_u = \begin{bmatrix} 0 & 0 & 0 & 0 & 0 \\ 0 & 0 & 1 & 0 & 0 \\ 0 & -(2^2) & 0 & 0 & 0 \\ 0 & 0 & 0 & 0 & 1 \\ 0 & 0 & 0 & -(4^2) & 0 \end{bmatrix} \hat{z}_u(t) \\ \hat{u}(t) = \Theta_u \hat{z}_u = \begin{bmatrix} 1 & 1 & 1 & 1 & 1 \end{bmatrix} \hat{z}_u \end{cases} \tag{5.38}$$

which corresponds to a step function and two sine-cosine pairs at the appropriate waveforms of 2 and 4 rad/s. The form of Eq. (5.38) suggests that the explicit frequency content in $u(t)$ must be known. However, F_u could generate additional waveforms and the estimator will still converge the error in both the internal state and unknown input. Equation (5.38) is a minimal representation of the input generator for the unknown input in Eq. (5.37) but is not the only valid internal model of the unknown input.

γ_e, which is a tunable parameter, is set to 1. Empirically, we have found the estimator performs well with moderate values of $\gamma_e \approx 1$. Note that increased values of γ_e will make the estimator increasingly nonlinear and more likely to diverge in the presence of discontinuous disturbances. On the other hand, lower values of γ_e make the estimator behave more like a non-adaptive estimator, and the correct form of A may not be recovered from A_m. The static gains K_x and K_u are simultaneously set by solving the Linear Quadratic Regulator problem for the composite system using a cheap control strategy. Figure 5.3 shows that the internal state error e_x converges to zero as expected. More importantly, Fig. 5.4 shows that the input estimator does converge onto the unknown input as expected. Finally, although the proof above does not provide it, L does converge to L_* in this specific case. When L does converge to L_*, the physical structure of the plant is recovered.

Of course, in this simple example, our input generator contained the exact basis functions present in the unknown input. In true application, it may be difficult to determine the waveform of the unknown input so precisely. This is especially true for processes that are not ergodic, as a spanning set of basis functions for a high dimensional space may be intractable. It is, therefore, worth considering the case where the input generator does not span the function space covered by the actual unknown input. We consider the exact same

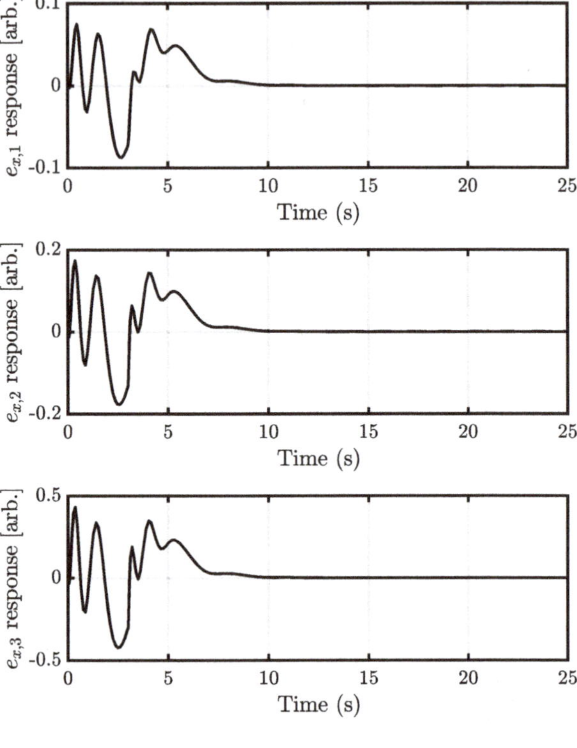

Fig. 5.3 Internal state error for adaptive unknown input observer. (Reproduced from [23])

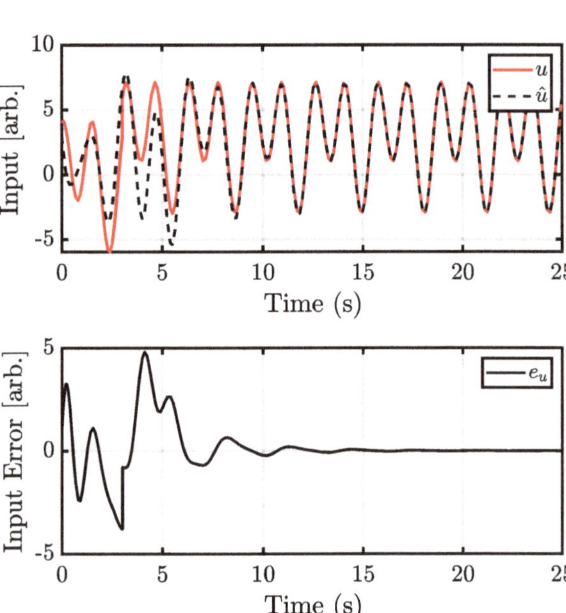

Fig. 5.4 Unknown input error for adaptive unknown input observer. (Reproduced from [23])

problem as above in Eqs. (5.34) and (5.36), but the unknown input Eq. (5.37) is modified as

$$u_2(t) = u_1(t) + c_4 \cos(8t). \tag{5.39}$$

Although the unknown input now contains a fourth waveform at a new frequency, the input generator Eq. (5.38) is not augmented with the waveform.

Figure 5.5 shows that without the inclusion of the fourth waveform, the estimator is no longer able to converge the internal error to zero. Rather, it reaches a neighborhood around zero. We note that the oscillation in the internal error has the same frequency as the fourth waveform. Further, when the unaccounted waveform is included, L does not converge to L_*. Because L does not converge, the correct physical structure of the plant is not recovered. Intuitively, the architecture is estimating the best fit unknown input given the available waveforms but is unable to represent the fourth waveform since it is linearly independent. Because the internal error does not converge to zero, but contains information related to the missing waveform, there is a future opportunity to augment the input generator online as a function of the estimator error. Further, since the estimator converges to a neighborhood around zero, there is an obvious extension to robust analysis.

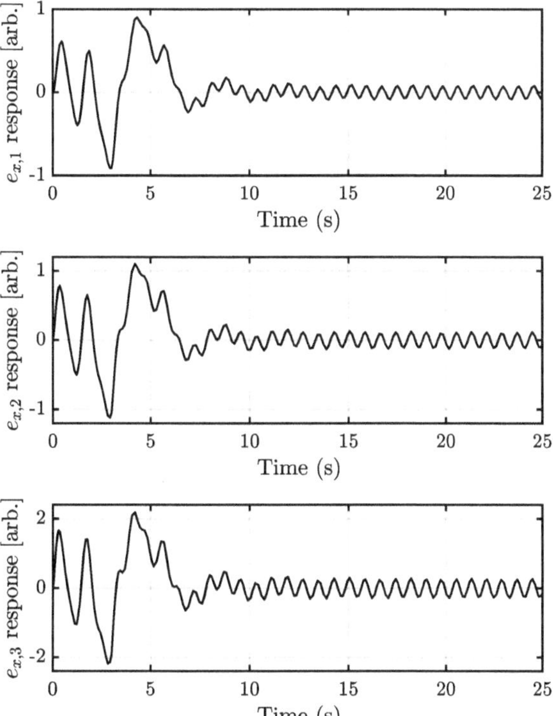

Fig. 5.5 Unknown input error for adaptive unknown input observer with excluded waveform. (Reproduced from [23])

5.6 Conclusions

We have provided a novel architecture for the simultaneous estimation of the internal state of a plant, and the system input in real time for models with uncertain dynamics. The architecture is crucially tied to an internal model of the unknown input, which must span the function space of the unknown input for reliable results. If the system of interest is ASD and has a bounded output, convergence of the state error and unknown input error is guaranteed, and the physical structure of the dynamics may be recovered.

While the architecture presented here does not treat general process disturbances or sensor noise, that extension was considered and the proof of convergence is presented in [35]. Further, known nonlinear plant dynamics were also treated in an additional proof in [36].

With this estimator architecture developed, we turn to its application on EEG brain wave modes in Chap. 6. *We will show that the specific formulation of this novel estimator is well suited to use on EEG data, because it treats the linear, nonlinear, and stochastic effects of the brain wave plant.*

References

1. A.S.C.T. Staff, A. Gelb, A.S. Corporation, *Applied Optimal Estimation* (MIT Press, 1974), https://books.google.com/books?id=KlFrn8lpPP0C
2. J. Hespanha, *Linear Systems Theory: Second Edition* (Princeton University Press, 2018), https://books.google.com/books?id=eDpDDwAAQBAJ
3. J. Na, A.S. Chen, G. Herrmann, R. Burke, C. Brace, Vehicle engine torque estimation via unknown input observer and adaptive parameter estimation. IEEE Trans. Veh. Technol. **67**(1), 409–422 (2018)
4. V. Gehlot, M.J. Balas, An evolving systems approach to the stable operation of dynamic formations and swarms of autonomous vehicles in a disruptive environment, in *2018 AIAA SPACE and Astronautics Forum and Exposition* (2018), p. 5272. https://arc.aiaa.org/doi/abs/10.2514/6.2018-5272
5. K. Tangsali, V.R. Krishnamurthy, Z. Hasnain, Generalizability of convolutional encoder–decoder networks for aerodynamic flow-field prediction across geometric and physical-fluidic variations. J. Mech. Des. **143**(5), 051704, 11 (2020). https://doi.org/10.1115/1.4048221
6. Y. Guan, M. Saif, A novel approach to the design of unknown input observers. IEEE Trans. Autom. Control **36**(5), 632–635 (1991)
7. J. Chen, R.J. Patton, H.-Y. Zhang, Design of unknown input observers and robust fault detection filters. Int. J. Control **63**(1), 85–105 (1996). https://doi.org/10.1080/00207179608921833
8. A. Fattouh, O. Sename, J. Dion, An unknown input observer design for linear time-delay systems, in *Proceedings of the 38th IEEE Conference on Decision and Control (Cat. No.99CH36304)*, vol. 4 (1999), pp. 4222–4227
9. M. Darouach, On the novel approach to the design of unknown input observers. IEEE Trans. Autom. Control **39**(3), 698–699 (1994)
10. B. Alenezi, J. Hu, S.H. Żak, Adaptive unknown input and state observers, in *2019 American Control Conference (ACC)* (2019), pp. 2434–2439

11. C. Johnson, Theory of disturbance-accommodating controllers, in *Control and Dynamic Systems*, vol. 12, ed. by C. Leondes (Academic Press, 1976), pp. 387–489, https://www.sciencedirect.com/science/article/pii/B9780120127122500135

12. M.J. Balas, Active control of persistent disturbances in large precision aerospace structures, in *Advances in Optical Structure Systems*, vol. 1303, ed. by J.A. Breakwell, V.L. Genberg, G.C. Krumweide, International Society for Optics and Photonics (SPIE, 1990), pp. 360–370. https://doi.org/10.1117/12.21519

13. J.H. Laks, L.Y. Pao, A.D. Wright, Control of wind turbines: past, present, and future, in *2009 American Control Conference* (2009), pp. 2096–2103

14. S. Torres, E. Mehiel, Nonlinear direct adaptive control and disturbance rejection for spacecraft, in *AIAA Guidance, Navigation, and Control Conference and Exhibit* (2006), p. 6038

15. M.J. Balas, S.A. Frost, Distributed parameter direct adaptive control using a new version of the Barbalat-Lyapunov stability result in Hilbert space, in *AIAA Guidance, Navigation, and Control (GNC) Conference* (2013), p. 4518

16. J. Zarei, E. Shokri, Robust sensor fault detection based on nonlinear unknown input observer. Measurement **48**, 355–367 (2014), https://www.sciencedirect.com/science/article/pii/S0263224113005587

17. F. Xu, J. Tan, X. Wang, V. Puig, B. Liang, B. Yuan, H. Liu, Generalized set-theoretic unknown input observer for LPV systems with application to state estimation and robust fault detection. Int. J. Robust Nonlinear Control **27**(17), 3812–3832 (2017), https://onlinelibrary.wiley.com/doi/abs/10.1002/rnc.3773

18. F. Xu, J. Tan, X. Wang, V. Puig, B. Liang, B. Yuan, Mixed active/passive robust fault detection and isolation using set-theoretic unknown input observers. IEEE Trans. Autom. Sci. Eng. **15**(2), 863–871 (2018)

19. M. Zhong, T. Xue, S.X. Ding, A survey on model-based fault diagnosis for linear discrete time-varying systems. Neurocomputing **306**, 51–60 (2018), https://www.sciencedirect.com/science/article/pii/S0925231218304715

20. S. Wang, J. Na, X. Ren, H. Yu, J. Yu, Unknown input observer-based robust adaptive funnel motion control for nonlinear servomechanisms. Int. J. Robust Nonlinear Control **28**(18), 6163–6179 (2018), https://onlinelibrary.wiley.com/doi/abs/10.1002/rnc.4368

21. D. Wang, K.-Y. Lum, Adaptive unknown input observer approach for aircraft actuator fault detection and isolation. Int. J. Adapt. Control Signal Process. **21**(1), 31–48 (2007), https://onlinelibrary.wiley.com/doi/abs/10.1002/acs.936

22. L. Ljung, *System Identification: Theory for the User* (Pearson Education, 1998), https://books.google.com/books?id=fYSrk4wDKPsC

23. T.D. Griffith, M.J. Balas, An adaptive control framework for unknown input estimation, *ASME International Mechanical Engineering Congress and Exposition* (2021)

24. M. Balas, R. Fuentes, A non-orthogonal projection approach to characterization of almost positive real systems with an application to adaptive control, in *Proceedings of the 2004 American Control Conference*, vol. 2 (2004), pp. 1911–1916

25. P.J. Davis, *Interpolation and Approximation* (Courier Corporation, 1975)

26. M.J.D. Powell, *Approximation Theory and Methods* (Cambridge University Press, 1981)

27. T.J. Rivlin, *An Introduction to the Approximation of Functions* (Courier Corporation, 1981)

28. G. Teschl, *Ordinary Differential Equations and Dynamical Systems*, vol. 140 (American Mathematical Society, 2012)

29. N.-H. Abel, Précis d'une theorie des fonctions elliptiques. J. für Die Reine Und Angew. Math. **1829**(4), 236–277 (1829)

30. K.S. Thapa Magar, M.J. Balas, S.A. Frost, Adaptive disturbance tracking control with wind speed reduced order state estimation for region ii control of large wind turbines, in *Smart Mate-*

rials, Adaptive Structures and Intelligent Systems, vol. 45097 (American Society of Mechanical Engineers, 2012), pp. 329–333

31. R.J. Fuentes, M.J. Balas, Robust model reference adaptive control with disturbance rejection, in *Proceedings of the 2002 American Control Conference (IEEE Cat. No.CH37301)*, vol. 5 (2002), pp. 4003–4008

32. C.E. Shannon, A mathematical theory of communication. ACM SIGMOBILE Mob. Comput. Commun. Rev. **5**(1), 3–55 (2001)

33. R. Ash, Information theory, in *Dover Books on Advanced Mathematics* (Dover Publications, 1990), https://books.google.com/books?id=yZ1JZA6Wo6YC

34. C. Johnson, Effective techniques for the identification and accommodation of disturbances, in *Proceedings of the 3rd Annual NASA/DOD Controls-Structures Interaction (CSI) Technical Conference* (1989), p. 163

35. T. Griffith, V.P. Gehlot, M.J. Balas, *Robust Adaptive Unknown Input Estimation with Uncertain System Realization*, https://arc.aiaa.org/doi/abs/10.2514/6.2022-0611

36. T.D. Griffith, M.J. Balas, Adaptive estimation of unknown inputs with weakly nonlinear dynamics, in *2022 American Control Conference (ACC)* (2022)

6.1   Introduction and Motivation

The UIO architecture introduced in Chap. 5 is guaranteed to converge its error to zero when there is no process disturbance in the output or plant dynamics. While this is a good starting point, we know from the discussion in Chap. 2 that EEG measures are very noisy. Therefore, additional complexity is needed to treat an unknown input estimator in the presence of noisy measurements. Here we introduce the necessary theoretical considerations for such a robust unknown input estimator, before proceeding to the analysis of the results on brain wave recordings.

State-space estimation of EEG data is appealing because of the wealth of existing knowledge associated with optimal state-space estimation. However, EEG signals are known to exhibit nonstationary effects which make modeling considerations difficult, especially when linear relationships form the basic structure of the model. As a result, many analytical tools which rely on stationary signals, such as the Fourier transform, require sliding windows or other compensations for the nonstationary dynamics. EEG signals have structured linear and nonlinear behavior, as well as general process noise arising from biological and artificial sources [1].

Here, we will demonstrate a novel state-space estimator, which addresses many of the modeling concerns when dealing with EEG data. This approach, which leverages an adaptive gain law, is highly nonlinear. The architecture presented performs three key functions. First, the adaptive law updates the model as the EEG data exhibits nonstationary effects. Second, this estimator estimates the exogenous input to the system, which improves the estimate by

---

* Reprinted with permission from "An adaptive unknown input approach to brain wave EEG estimation," by T. Griffith, V. Gehlot, M. Balas, J. Hubbard, 2021. *Biomedical Signal Processing and Control* Vol. 79, pp. 104083, Copyright 2022 by Elsevier.

accounting for dynamics that the model may not generate internally. Finally, it is robust to general modeling error and process uncertainty.

The estimation of the exogenous input is particularly unique here and offers additional analytical insight into brain wave EEG recordings. Others have leveraged state-space estimation techniques for the analysis of brain waves. Wong et al. [2] demonstrated an effective modeling procedure, accounting for the nonstationary dynamics by varying the driving noise to the model. Cheung et al. [3] leveraged a more general framework to model cortical connectivity from EEG data. This work was also confronted with the uncertainty associated with spatial dynamics, which we will attempt to further address here. More recently, [4] combined multiple state-space models with a variety of classification algorithms for successful sleep stage classification. Finally, [5] demonstrated a Bayesian state-space approach for the tracking of attentional states. In each of these cases, the appeal of state-space estimation is clear: real-time predictions backed by rigorous mathematical structure. Here, our approach is focused on maintaining a flexible state-space framework, with an eye toward interpretable modeling outcomes. This work is focused on the *modeling* of the EEG recording, and less so on a specific analytical outcome. We seek a widely applicable, real-time estimator which yields interpretable outcomes for future analysis.

6.2 Technical Approach

Spatio-temporal brain wave dynamics contain structured linear and nonlinear behavior, as well as unstructured random noise. The approach in this work is especially focused on elegantly treating each of these dynamic effects. As we will show in Sect. 6.3, the proposed approach operates exactly in this way, filtering the random noise, while estimating the linear and nonlinear dynamics. We treated the linear dynamics in Chap. 4. Ordinary differential equations (ODEs) can canonically represent a given linear system's dynamics in state space as

$$\dot{x}(t) = Ax(t) + Bu(t) + v_x \tag{6.1}$$

$$y(t) = Cx(t). \tag{6.2}$$

Internal plant states evolve in the n-dimensional vector space $X \equiv \mathbb{C}^n$. The input is in the m-dimensional space $U \equiv \mathbb{C}^m$. The output is in the p-dimensional space $Y \equiv \mathbb{C}^p$. v_x is a signal that represents general uncertainty or noise in the system model. We include this term here so that it carries through our theoretical analysis, offering some insight into the effects of uncertainty on the estimation of EEG brain waves generally. We will show in Sect. 6.2.3 that the accuracy of our real-time estimate is greatly influenced by the bounded uncertainty v_x. y contains the measurements of a prescribed set of sensors as a time-variant vector. $u \in U$ is the input to the system. In a mechanical system, the inputs can often be precisely controlled for improved analysis. The unknown inputs act through the matrix B, which distributes each unknown input across the appropriate internal states.

6.2.1 Treating the Nonlinear Effects of Brain Waves

As introduced in Sect. 6.1, because brain waves exhibit nonlinear, nonstationary dynamics, a single model of the form Eq. (6.2) is unlikely to describe the global brain wave dynamics. Much of modern control theory is dedicated to analyzing and stabilizing nonlinear systems. Often, engineers linearize the plant dynamics around a discrete set of known operating conditions. Each linearization takes the form of Eq. (6.2), and then a variety of algorithms may switch between linearizations as needed to improve the model accuracy or performance outcomes [6–8]. Of course, it is difficult to assess which operating points are important to brain waves. Therefore, we introduce a novel adaptive estimation scheme to simultaneously treat nonlinear plant dynamics and estimate the unknown system input. This adaptive estimator updates the physical structure of (A, C) to better match the observed plant dynamics. The adaptive estimator is highly nonlinear and modifies (A, C) to account for parametric uncertainty and nonlinear effects. Simultaneously, the estimator forms a representation of the system input u which we call $\hat{u}$. We will hypothesize and then validate that the unknown system input acts evenly on the spatial brain wave dynamics, which implies that B is a matrix of ones. Here, we present this adaptive approach to real-time brain wave estimation as

$$\dot{\hat{x}} = \left(A_m + BL(t)C\right)\hat{x} + B\hat{u} + K_x e_y; \tag{6.3}$$

$$\hat{y} = C\hat{x} \tag{6.4}$$

$$\dot{L} = -e_y y^* \gamma_e - \alpha L; \ \gamma_e > 0, \ \alpha > 0. \tag{6.5}$$

Again, and crucially $A \neq A_m$. A, B, C, and K_x are all static matrices of appropriate size. Notice, in Chap. 5, the adaptive gain law was $\dot{L} = -e_y y^* \gamma_e$. Here, we have added an additional term $-\alpha L$. This term acts precisely as a first order filter on the adaptive gain, which prevents it from responding too suddenly to a perturbation which is noise. We may only use the observable information in the system (the output y) to correct A_m. Through adaptive output feedback, $A_m + BL(t)C$ approximates A, even when faced with nonlinear dynamics. In this view, the term $BL(t)C$ updates A_m to track the nonlinear brain wave dynamics. Each $L(t)$ yields a different linear model of the form Eq. (6.2). We will demonstrate that Eq. (6.3) generates estimates $\hat{y}$ of the measured brain wave signal y that are useful to cognitive modeling outcomes.

6.2.2 Treating the Unknown Input

In order to estimate an unknown input, we modify the approach in [9]. Identifying u directly from output data alone is an unsolved problem in state estimation and system identification. Here, we solve this issue by modeling the unknown input as a weighted linear combination of basis waveforms. We approximate the true input space U with an nth order linear combination

$$\hat{u} = \sum_{i=1}^{N} c_i f_i(t).$$

Now, rather than identifying u, we change the problem into identifying the c_i's which give the best prediction of the EEG data given the linear dynamics (A_m, C). The weighting coefficients c_i may be discontinuous and change with time. Accordingly, our estimation procedure should allow for this effect. To use this approach alongside the linear dynamic model in Eq. (6.3), we must formulate Eq. (6.6) in state-space terms

$$\dot{z}_u = F_u z_u(t) \tag{6.6}$$
$$\hat{u}(t) = \Theta_u z_u(t). \tag{6.7}$$

This set of coupled ordinary differential equations serves a very direct purpose. The matrix F_u generates the desired waveforms which we suspect the unknown input contains. For brain wave dynamics, it may be desirable to consider some oscillatory trigonometric functions, such as sines or cosines, since brain waves are known to have physically significant spectral behavior. We may also choose a polynomial basis or a set of natural logarithms. As long as the waveforms are causal and linearly independent from one another, they can be generated by Eq. (6.7). The matrix Θ_u simply serves to superpose the waveforms into a single estimate of $\hat{u}$. This approach is amenable to multiple unknown inputs as we demonstrated in [10]. Notice that while this approach makes the problem solvable, we have to be able to say something about the waveforms which makes up the unknown input; some additional insight is necessary to solve this problem.

As in our formulation of the plant dynamics in Eq. (6.2), we suppose there is also some bounded, unknown uncertainty associated with the unknown input. That is, we likely do not have a perfect formulation of Eq. (6.7), so the true unknown input takes the following form

$$\dot{z}_u = F z_u(t) + v_u \tag{6.8}$$
$$u(t) = \Theta_u z_u(t). \tag{6.9}$$

Because brain wave measures are not known to diverge, it is reasonable to bind these uncertainties for further analysis as

$$v = \begin{bmatrix} v_x \\ v_u \end{bmatrix} \tag{6.10}$$

$$||v|| \leq M_v \leq \infty. \tag{6.11}$$

Notice, this formulation treats many different uncertainties that are known to influence brain wave EEG models. First, the formulation of the state matrix A_m may not be especially accurate. We treat this effect with the adaptive parameter $L(t)$. Second, the input u is unknown. This effect is treated with the persistent input approach, imposing known waveforms on the unknown input. Further, there is a general perturbation from unknown sources v that we also analyze. Rather than simply lumping all the uncertainties into a single term, we structure each using information about the system to inform our modeling approach. Notice, this estimator is highly nonlinear. Here, we are leveraging the nonlinearity of the estimator to treat the nonstationarity of the EEG signals. The matrix A_m is accommodated in real time for the statistical shifts and is still amenable to modal decomposition as described in Sect. 2.3.1.

6.2.3 Estimator Architecture and Proof of Convergence

In combination, this adaptive unknown input estimator (UIO), takes the form in Fig. 6.1. We now proceed to the guarantee of convergence and stability for this approach. Because EEG waves are not known to diverge, it is especially important that our estimator also does not diverge. We begin with a statement of the final result in Theorem 1.

Theorem 1 (Robust Output Feedback Model Direct Reference Brain Wave Estimation) *Consider the nonlinear, coupled error system differential equations,*

$$\dot{e} = \bar{A}_c e + \bar{B}(L(t) - L_*)y + v \tag{6.12}$$

$$e_y = \bar{C}e \tag{6.13}$$

$$\dot{L}(t) = -e_y y^* \gamma_e - \alpha L(t) \tag{6.14}$$

We make use of the following assumptions:

(i)   the matrix triple $(\bar{A}_c, \bar{B}, \bar{C})$ is ASD,

(ii) $\exists M_k > 0 \ni \sqrt{tr(L^* L_*^*)} \leq M_k$,

(iii) $\exists M_v > 0 \ni \sup_{t \geq 0} ||v(t)|| \leq M_v$,

(iv) $\exists \alpha > 0 \ni \alpha \leq \frac{\lambda_{min}(\bar{Q})}{\lambda_{max}(\bar{P})}$, and

(v) γ_e satisfies $tr(\gamma_e^{-1}) \leq \left(\frac{M_v}{\alpha M_k}\right)^2$, where $\lambda_{max}(\bar{P})$, $\lambda_{min}(\bar{P})$ are the maximum and minimum eigenvalues of $\bar{P}$ and $\lambda_{min}(\bar{Q})$ is the minimum eigenvalue of $\bar{Q}$ with respect to Definition 1. Then, the matrix $L(t)$ is bounded and the error state $e(t)$ exponentially approaches the n-ball of radius

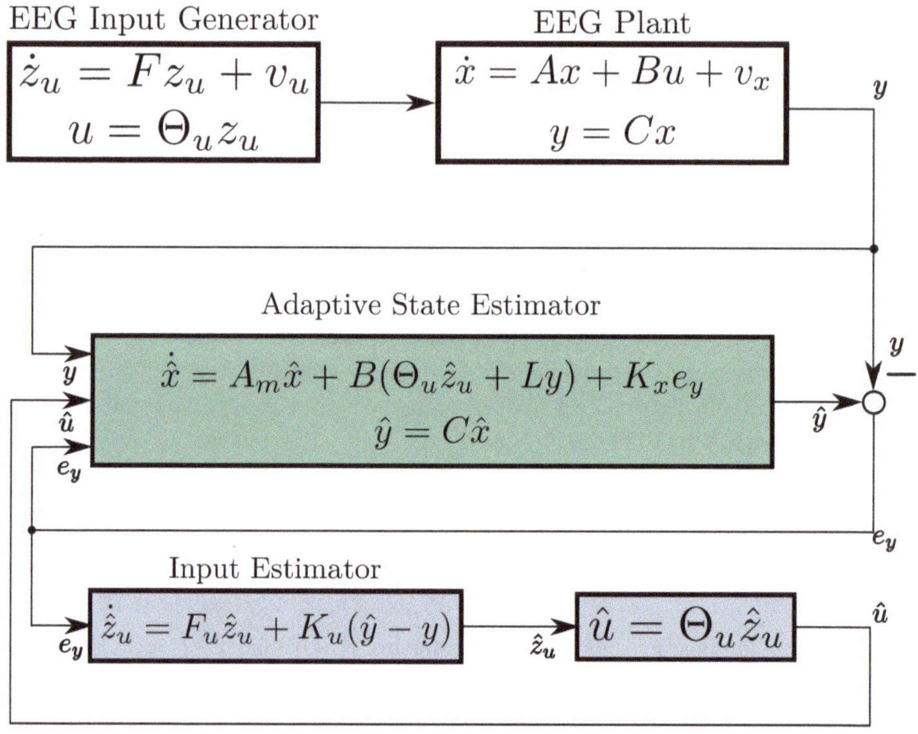

Fig. 6.1 Robust adaptive unknown input estimator for EEG brain wave dynamics. (Reproduced from [11])

$$\frac{\left(1 + \sqrt{\lambda_{max}(\bar{P})}\right)}{\alpha \sqrt{\lambda_{min}(\bar{P})}} M_v.$$

Note that although the result above guarantees the convergence of the error $e(t)$, it does not guarantee the convergence of $\Delta L = L(t) - L_*$. The adaptive gain $L(t)$ evolves freely and may reduce the performance of the estimator due to excessive gains. One possible solution is the inclusion of a projection operator to bound the maximum possible gain $L(t)$ as in [12].

Definition 1 Almost Strict Dissipativity

The following convergence results rely greatly on assumption (*i*), the notion of almost strict dissipativity (ASD). The system in Eq. (6.2) is ASD if it meets the Kalman-Yakubovich conditions:

$$A_c^* P + P A_c = -Q \tag{6.15}$$

$$P B = C^* \tag{6.16}$$

for some positive definite matrices P and Q. This matrix $A_c = A + B G^* C$ may stabilize A with a stable gain matrix G and is equivalent to the closed loop matrix for the system in Eq. (6.2) under the feedback law $u = G y$.

Given the novel scheme in Fig. 6.1, the composite error dynamics $\bar{x} = \begin{bmatrix} x \\ z_u \end{bmatrix}$ are

$$\dot{e} = (\bar{A} + \bar{K}\bar{C})e + \bar{B}w + v \tag{6.17a}$$

$$\begin{bmatrix} \dot{e}_x \\ \dot{e}_z \end{bmatrix} = \left(\begin{bmatrix} A_m & B\Theta_u \\ 0 & F_u \end{bmatrix} + \begin{bmatrix} K_x \\ K_u \end{bmatrix} \begin{bmatrix} C & 0 \end{bmatrix} \right) \begin{bmatrix} e_x \\ e_z \end{bmatrix}$$

$$+ \begin{bmatrix} B \\ 0 \end{bmatrix} w + \begin{bmatrix} v_x \\ v_u \end{bmatrix} \tag{6.17b}$$

$$= \underbrace{\begin{bmatrix} A_m + K_x C & B\Theta_u \\ K_u C & F_u \end{bmatrix}}_{\bar{A}_c} \begin{bmatrix} e_x \\ e_z \end{bmatrix}$$

$$+ \begin{bmatrix} B \\ 0 \end{bmatrix} w + \begin{bmatrix} v_x \\ v_u \end{bmatrix} \tag{6.17c}$$

$$e_y = \begin{bmatrix} C & 0 \end{bmatrix} \begin{bmatrix} e_x \\ e_z \end{bmatrix}.$$

The convergence results for these error dynamics stem from the consideration of the Lyapunov function considering both the error e and the adaptive gain ΔL. Consider a Lyapunov function for the error e as

$$\frac{\lambda_{\min}(\bar{P})}{2}||e||^2 \leq V_1(e) \equiv \frac{1}{2}e^* \bar{P} e \leq \frac{\lambda_{\max}(\bar{P})}{2}||e||^2 \tag{6.18}$$

where $\bar{P}$ is the appropriate symmetric positive matrix satisfying Definition 1. $\lambda_{\max}(\bullet)$ is the largest eigenvalue of the argument. Correspondingly, $\lambda_{\min}(\bullet)$ is the smallest eigenvalue. The time derivative of V_1 is

$$\dot{V}_1(e) = e^* \bar{P} \dot{e} \tag{6.19a}$$

$$= e^* \bar{P}(\bar{A}_c e + \bar{B}w + v) \tag{6.19b}$$

$$= e^* \bar{P} \bar{A}_c e + e^* \bar{P} \bar{B}w + e^* \bar{P}v \tag{6.19c}$$

$$= \frac{1}{2}e^*(\bar{P}\bar{A}_c + \bar{A}_c^* \bar{P})e + e^* \bar{P} \bar{B}w + e^* \bar{P}v. \tag{6.19d}$$

Using the Kalman-Yakubovich conditions in Definition 1, $\dot{V}_1$ is reduced to

$$\dot{V}_1(e) = -\frac{1}{2}e^*\bar{Q}e + e^*\bar{C}^*w + e^*\bar{P}v \tag{6.20a}$$

$$= -\frac{1}{2}e^*\bar{Q}e + (e_y, w) + (\bar{P}e, v). \tag{6.20b}$$

The adaptive gain ΔL may also cause instability. We introduce a second Lyapunov function for the adaptive gain using $tr(\bullet)$, the matrix trace operator

$$V_2(\Delta L) \equiv \frac{1}{2}tr(\Delta L\gamma_e^{-1}\Delta L^*). \tag{6.21}$$

γ_e is a tunable parameter which is also symmetric positive definite. The time derivative of V_2 is

$$\dot{V}_2(\Delta L) = tr(\Delta\dot{L}\gamma_e^{-1}\Delta L^*) \tag{6.22a}$$

$$= tr\big((-e_y y^*\gamma_e - \alpha L)\gamma_e^{-1}\Delta L^*\big) \tag{6.22b}$$

$$= tr(-e_y y^*\Delta L^* - \alpha L\gamma_e^{-1}\Delta L^*) \tag{6.22c}$$

$$= tr(-e_y y^*\Delta L^*) + tr(-\alpha L\gamma_e^{-1}\Delta L^*) \tag{6.22d}$$

$$= -(e_y, w) + tr(-\alpha L\gamma_e^{-1}\Delta L^*). \tag{6.22e}$$

In order to prove the stability of the state error e_x, the input error e_z, and the adaptive gain ΔL, the Lyapunov functions are combined into a composite Lyapunov function for further analysis. The composite function is defined as

$$V(e, \Delta L) \equiv V_1(e) + V_2(\Delta L) \tag{6.23}$$

$$= \frac{1}{2}e^*\bar{P}e + \frac{1}{2}tr(\Delta L\gamma_e^{-1}\Delta L^*). \tag{6.24}$$

The time derivative of the composite Lyapunov function is therefore

$$\dot{V}(e, \Delta L) = \dot{V}_1(e) + \dot{V}_2(\Delta L) \tag{6.25a}$$

$$= -\frac{1}{2}e^*\bar{Q}e + (e_y, w) + (\bar{P}e, v)$$
$$\quad - (e_y, w) + tr(-\alpha L\gamma_e^{-1}\Delta L^*) \tag{6.25b}$$

$$= -\frac{1}{2}e^*\bar{Q}e + (\bar{P}e, v)$$
$$\quad + tr(-\alpha L\gamma_e^{-1}\Delta L^*) \tag{6.25c}$$

$$= -\frac{1}{2}e^*\bar{Q}e + (\bar{P}e, v)$$
$$\quad + tr(-\alpha(\Delta L + L_*)\gamma_e^{-1}\Delta L^*). \tag{6.25d}$$

Using Sylvester's inequality yields

$$\dot{V}(e, \Delta L) \leq -\lambda_{\min}(\bar{Q})||e||^2 - \alpha tr(\Delta L \gamma_e^{-1} \Delta L^*)$$
$$+ \alpha |tr(L_* \gamma_e^{-1} \Delta L^*)| + |(\bar{P}e, v)|. \tag{6.26}$$

We make use of assumption *(vi)* resulting in

$$\dot{V}(e, \Delta L) \leq -\lambda_{\min}(\bar{Q})||e||^2 - \alpha tr(\Delta L \gamma_e^{-1} \Delta L^*)$$
$$+ \alpha |tr(L_* \gamma_e^{-1} \Delta L^*)| + |(\bar{P}e, v)| \tag{6.27a}$$
$$\leq -\lambda_{\min}(\bar{Q})||e||^2 - \alpha tr(\Delta L \gamma_e^{-1} \Delta L^*)$$
$$+ \alpha |tr(L_* \gamma_e^{-1} \Delta L^*)| + |(\bar{P}e, v)|. \tag{6.27b}$$

Equation (6.18) can be rewritten in terms of the error e, considering only the upper bound

$$\frac{1}{\lambda_{\max}(\bar{P})}(e^* \bar{P}e) \leq ||e||^2. \tag{6.28}$$

Substitute Eq. (6.28) into Eq. (6.27b)

$$\dot{V}(e, \Delta L) \leq -\left(\frac{\lambda_{\min}(\bar{Q})}{\lambda_{\max}(\bar{P})}\right) e^* \bar{P}e$$
$$- \alpha tr(\Delta L \gamma_e^{-1} \Delta L^*)$$
$$+ \alpha |tr(L_* \gamma_e^{-1} \Delta L^*)| + |(\bar{P}e, v)|. \tag{6.29}$$

We make use of assumption *(iv)* to obtain

$$\dot{V}(e, \Delta L) \leq -\alpha e^* \bar{P}e - \alpha tr(\Delta L \gamma_e^{-1} \Delta L^*)$$
$$+ \alpha |tr(L_* \gamma_e^{-1} \Delta L^*)| + |(\bar{P}e, v)|$$
$$\leq -\alpha \underbrace{(e^* \bar{P}e + tr(\Delta L \gamma_e^{-1} \Delta L^*))}_{2V}$$
$$+ \alpha |tr(L_* \gamma_e^{-1} \Delta L^*)| + |(\bar{P}e, v)| \tag{6.30a}$$
$$\leq -2\alpha V + \alpha |tr(L_* \gamma_e^{-1} \Delta L^*)|$$
$$+ |(\bar{P}e, v)|.$$

The Cauchy-Schwarz inequality is appropriate for the remaining trace operator

$$|tr(L_* \gamma_e^{-1} \Delta L^*)| \leq ||L_*||_2 ||\Delta L||_2 \tag{6.31}$$

and inner product

$$|(\bar{P}e, v)| \leq ||\bar{P}^{\frac{1}{2}}v|| \, ||\bar{P}^{\frac{1}{2}}e|| = \sqrt{(\bar{P}v, v)}\sqrt{(\bar{P}e, e)}. \tag{6.32}$$

Insert Eqs. (6.31) and (6.32) into Eq. (6.30a) for

$$\dot{V}(e, \Delta L) \leq -2\alpha V + \alpha ||L_*||_2 ||\Delta L||_2$$
$$+ \sqrt{(\bar{P}v, v)}\sqrt{(\bar{P}e, e)} \tag{6.33a}$$

$$\dot{V} + 2\alpha V \leq \alpha ||L_*||_2 \underbrace{||\Delta L||_2}_{\leq \sqrt{2}V^{1/2}}$$
$$+ \sqrt{(\bar{P}v, v)} \underbrace{\sqrt{(\bar{P}e, e)}}_{\leq \sqrt{2}V^{1/2}} \tag{6.33b}$$

$$\dot{V} + 2\alpha V \leq \left(\alpha ||L_*||_2 + \sqrt{(\bar{P}v, v)}\right)\sqrt{2}V^{\frac{1}{2}}. \tag{6.33c}$$

We arrive at

$$\frac{\dot{V} + 2\alpha V}{V^{\frac{1}{2}}} \leq \sqrt{2}\left(\alpha ||L_*||_2 + \sqrt{(\bar{P}v, v)}\right). \tag{6.34}$$

Now, we treat the $||L_*||_2$ term

$$||L_*||_2 \equiv [tr(L_* \gamma_e^{-1} L_*^*)]^{\frac{1}{2}}. \tag{6.35}$$

The trace operator is invariant under cyclic permutations

$$||L_*||_2 = [tr(L_* L_*^* \gamma_e^{-1})]^{\frac{1}{2}}. \tag{6.36}$$

Further, because $tr(AB^*) \leq tr(AA^*)^{\frac{1}{2}} tr(BB^*)^{\frac{1}{2}}$,

$$||L_*||_2 \leq \left[(tr(L_*^* L_* L_* L_*^*)^{\frac{1}{2}} tr(\gamma_e^{-1} \gamma_e^{-1})^{\frac{1}{2}}\right]^{\frac{1}{2}}$$
$$= [tr(L_* L_*^*)]^{\frac{1}{2}} [tr(\gamma_e^{-1})]^{\frac{1}{2}}. \tag{6.37}$$

We make use of assumptions *(ii)* and *(vii)* resulting in

$$||L_*||_2 \leq (M_k)\left(\frac{M_v}{\alpha M_k}\right) = \frac{M_v}{\alpha}. \tag{6.38}$$

Substituting Eq. (6.38) into Eq. (6.34) yields

$$\frac{\dot{V} + 2\alpha V}{V^{\frac{1}{2}}} \leq \sqrt{2}\left(M_v + \sqrt{(\bar{P}v, v)}\right) \tag{6.39a}$$

$$\leq \sqrt{2}\left(M_v + \sqrt{\lambda_{\max}(\bar{P})}M_v\right) \tag{6.39b}$$

$$\leq \sqrt{2}\left(1 + \sqrt{\lambda_{\max}(\bar{P})}\right)M_v. \tag{6.39c}$$

Notice that

$$\frac{d}{dt}(2e^{\alpha t}V^{\frac{1}{2}}) = e^{\alpha t}\frac{\dot{V} + 2\alpha V}{V^{\frac{1}{2}}}$$
$$\leq \sqrt{2}\left(1 + \sqrt{\lambda_{\max}(\bar{P})}\right)M_v. \tag{6.40}$$

Equation (6.40) can be integrated from 0 to τ resulting in

$$e^{\alpha \tau}V(\tau)^{\frac{1}{2}} - V(0)^{\frac{1}{2}}$$
$$\leq \frac{(1 + \sqrt{\lambda_{\max}(\bar{P})})M_v}{\alpha}(e^{\alpha t} - 1). \tag{6.41}$$

This expression can be manipulated to yield

$$V(\tau)^{\frac{1}{2}} \leq V(0)^{\frac{1}{2}}e^{-\alpha t}$$
$$+ \frac{(1 + \sqrt{\lambda_{\max}(\bar{P})})M_v}{\alpha}(1 - e^{-\alpha t}). \tag{6.42}$$

Because $V(\tau)$ is a function of the state error e and the adaptive gain matrix L, the boundedness of $V(\tau)$ bounds e and L. Equation (6.42) bounds $V(\tau)^{\frac{1}{2}}$, so $V(\tau)$ is necessarily bounded also. The use of Sylvester's inequality further yields

$$\sqrt{\lambda_{\max}(\bar{P})}||e|| \leq V(t)^{1/2}. \tag{6.43}$$

Equation (6.43) is substituted into Eq. (6.42) and we take the limit superior for the final result

$$\limsup_{t \to \infty}||e(t)|| \leq \frac{\left(1 + \sqrt{\lambda_{\max}(\bar{P})}\right)}{\alpha\sqrt{\lambda_{\min}(\bar{P})}}M_v \equiv R_*. \tag{6.44}$$

which exponentially bounds the state error e. As a result, given the assumptions in Theorem 1, the adaptive gain is bounded and the error dynamics converge to a ball of radius R_*. The proof is complete.

6.2.4 Datasets

In the following section, we will demonstrate the efficacy of this approach on a representative dataset. As discussed in Sect. 6.1, the focus of this work is modeling, not analysis. Accordingly, the dataset presented here is selected for its rigor in collection and preprocessing. The Database for Emotion Analysis using Physiological Signals (DEAP) [13] was selected for the results that follow. The database consists of 40 distinct 60 s long trial videos for each of the 32 different subjects. The 60 s trials are designed to elicit specific emotions from the subjects, but that analysis is not focused on here. The data was collected according to the

10–20 system, with 32 total channels of EEG data averaged to the common reference. This means that our plant output y is a 32 by 1 column vector. For the purposes of this work, no additional preprocessing was added to the data.

Additionally, the database accompanying [14] was considered as a validation data set for the analysis in Sect. 6.3.3. This open source data set is comprised of EEG recordings for 25 different subjects using the Emotiv Epoc+ device, which has 14 electrodes. In this database, subjects are shown e-commerce products from 14 different categories and asked to rate the product into binary categories of like or dislike. The provided EEG signals are unprocessed, so for the purposes of consistency, the same steps were done to this data as in the DEAP data.

6.3   Results

Having established the technical background for this adaptive method of estimating EEG dynamics in real time, we will demonstrate the efficacy of this approach on the EEG data introduced in Sect. 6.2.4. Because of the nonstationary nature of EEG measures, we do not expect the estimator error to converge for extended periods of time per Eq. (6.44). To validate our approach, we will compare the adaptive UIO in Eq. (6.3) with simpler estimators. We hypothesize that the adaptive UIO will outperform the simpler estimators. First, if the adaptive scheme and unknown input estimator were removed from Eq. (6.3), the estimator would take a simplified form

$$\dot{\hat{x}}_k = A_m \hat{x}_k + K_k e_y; \tag{6.45}$$
$$\hat{y} = C \hat{x}_k, \tag{6.46}$$

which is a Kalman filter. The estimator in Eq. (6.45) does not account for the input nor the parametric uncertainty in A_m, but it has been widely and successfully implemented in a variety of applications and so is a good point of comparison [15]. We could reintroduce the adaptive gain law alone, without treating the unknown input

$$\dot{\hat{x}} = \left(A_m + BL(t)C\right)\hat{x} + K_a e_y; \tag{6.47}$$
$$\hat{y} = C \hat{x}, \tag{6.48}$$

which we will denote as the adaptive only estimator. Third, we could estimate the unknown input without accounting for the uncertainty in A_m. This estimator takes the form

$$\dot{\hat{x}} = A_m \hat{x} + B \hat{u} + K_I e_y; \tag{6.49}$$
$$\hat{y} = C \hat{x}. \tag{6.50}$$

As for the static gains matrices, K_k, K_a, and K_I are set via LQR methods with $Q = I$, $R = I$, and $N = 0$. This is the same cost function, though of a different dimension, as the one used to set the static gains in Eq. (6.3). We will compare our novel UIO with the Kalman filter, adaptive only estimator, and input only estimator. As a reminder, we are putting aside the formulation of (A_m, C), as that process was developed previously in [16], but know that it is a best-fit estimate of the linear dynamics without knowledge of the input.

6.3.1 Performance Benefits of UIO

Initially, we may be interested in the immediate comparison of the relative error between our UIO and the other estimators. Figure 6.2 shows the 2-norm of the error $||e_y(t)||_2$ across all subjects and trials in the DEAP database. That is, a matrix pair (A_m, C) is determined using the procedure in [16] from a single trial of 60 s EEG data in the dataset per subject. The estimators in Eqs. (6.45) and (6.3) are then used to predict the EEG signal of the remaining 60 s trials. For each of the 39 trials, an error $e_y = \hat{y} - y$ is obtained as a time series with the same dimension as y. The norm of each error $e_y(t)$ is taken to give a single measure of the error per trial. As a result, for 32 subjects with 40 trials each, 1248 normed errors are obtained for each estimator, representing the accuracy of each estimator on unmodeled data. For the purposes of comparison, γ_e was static across all trials, even though some of the trials may need more or less influence from the adaptation of A_m. The box plot for these errors is presented in Fig. 6.2 as a method for comparing the relative performance of each estimator. Note that on average, our adaptive UIO yields an error that is 26% smaller than the error of the standard Kalman filter. That is, by accounting for the nonlinear effects and exogenous input to the brain wave plant, the estimate improves as a whole. Even for the least accurate case here, the adaptive UIO outperforms the best Kalman filter estimate. Of course, it can also be seen that the adaptive scheme has a wider distribution; the "goodness" of the estimate varies more. This effect is also seen in the performance of estimator Eq. (6.49), so we attribute the spread to the inclusion of $\hat{u}$. This effect may be mitigated by the possible optimization of the unknown input basis functions [17]. It remains a curious result of this approach which requires more exploration in the future. Of course, for a fair comparison, the same LQR gains and γ_e are selected for all subjects in the database. Individually tuning the gains and γ_e matrix for each case would improve the error spread.

Overall, we conclude that this novel UIO outperforms the Kalman filtering approach by accounting for the unknown input and nonlinear effects in the true plant. Using the estimator error e_y as a metric for comparison is rigorously useful, but obscures some of the other benefits of the adaptive UIO. It's promising that the adaptive unknown input estimator is more accurate than simpler estimators, but there are additional analytical benefits to this approach also.

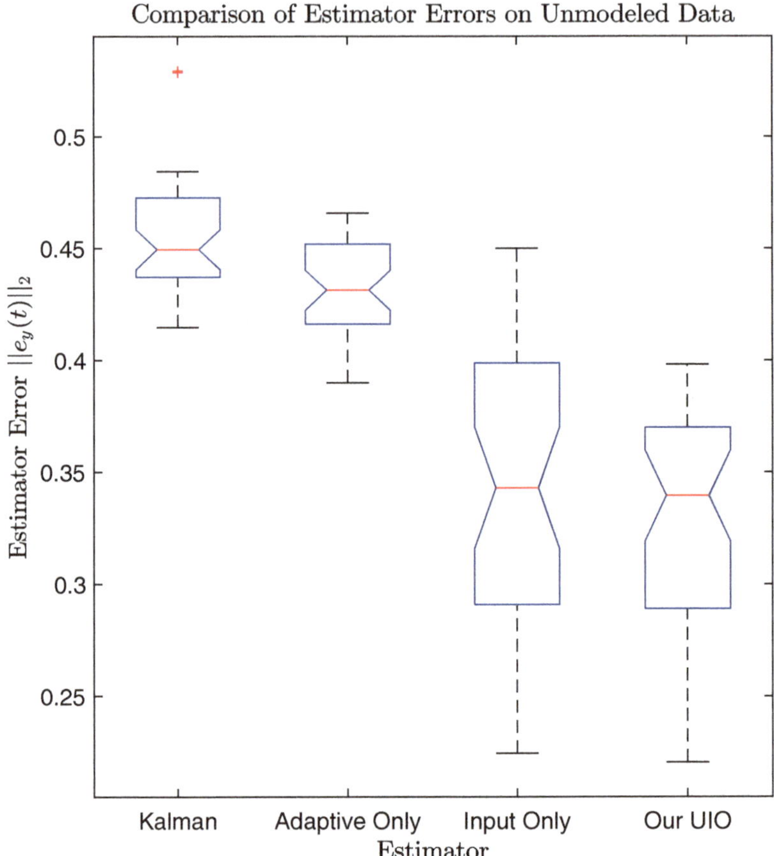

Fig. 6.2 Comparison of estimator error for unmodeled data across proposed estimators. While our adaptive UIO has an increased standard error, it outperforms Kalman filtering in all cases and is better on average than the other estimators. (Reproduced from [11])

6.3.2 Analytical Benefits of UIO

The most obvious analytical benefit of this approach is the estimate of the input. It has been difficult historically to translate common stimuli to "brain wave space". Much discussion in Sect. 6.2.2 was dedicated to the formulation of the input waveform generator. For effective modeling, we must select the right cardinality and structure for the input basis. For example, many previous applications have used four polynomial functions ($[1, t, t^2, t^3]$) with only moderate improvement from higher order terms [18]. Here, we will consider this heuristic basis set as a baseline for comparison. For brain wave EEG modeling, we may choose to use sine and cosine waveforms because EEG measures are known to have oscillatory behavior. This approach yields more physical significance to the estimate of the unknown input. A single paired sine-cosine waveform may be generated with the following persistent input matrix

$$F_u = \begin{bmatrix} 0 & 1 \\ -\omega^2 & 0 \end{bmatrix}, \tag{6.51}$$

where ω is the angular frequency of the sine-cosine pair. Additional frequencies can be generated by stacking the matrices along the diagonal. For example, a sine-cosine pair at 2 and 5 rad/s may simultaneously be generated with

$$F_u = \begin{bmatrix} 0 & 1 & 0 & 0 \\ -2^2 & 0 & 0 & 0 \\ 0 & 0 & 0 & 1 \\ 0 & 0 & -5^2 & 0 \end{bmatrix}. \tag{6.52}$$

Since this particular EEG data was bandpass filtered between 4 and 45 Hz, we should expect the estimator accuracy not to improve with additional waveforms past 45 Hz. Figure 6.3 shows the accuracy of all trials as the estimator generates increasingly many waveforms at increasingly high frequencies. The solid horizontal line represents the baseline polynomial estimator. As expected, the accuracy does not improve when generating sine-cosine pairs faster than 45 Hz, and the best-performing sine-cosine basis set outperforms the standard polynomial basis. This reinforces the notion that the unknown input generator should be formulated using insights from the plant output. Because the data was filtered between 4–45 Hz, the best input estimate will come from an input generator that covers this spectrum. For brain wave models, we recommend sine and cosine waveforms for the input generator because of the observed accuracy and physical insight attached to this particular basis. Additionally, while the polynomial input generator is simpler than the sine-cosine input generator, the polynomial input generator is a very sparse matrix and can result in uncontrollability of the dynamics in Eq. (6.17a).

With a suitable input generator established, analysis of the unknown input estimate itself is possible. Figure 6.4 shows an example of the unknown input using an input generator of sine and cosine waveforms. As we might expect, the estimate of the unknown input resembles the source EEG data. This vector $\hat{u}$ improves the model efficacy and may be correlated with external stimuli. Modeling, not analysis, is the focus of this work, so we leave such a correlation for future efforts. As a reminder, the estimate of this waveform is generated in real time. EEG data flows into the estimator, and the estimator provides an estimate of the internal plant state $\hat{x}$ and an estimate of unknown input $\hat{u}$. The unknown input estimate also includes the individual weighting coefficients for each of the generated basis waveforms in Eq. (6.7).

6.3.2.1 L(t) Is a Measure of Nonlinearity

In addition to the estimate of the unknown input, the adaptive gain matrix $L(t)$ may also be useful in analysis. As a reminder, $L(t)$ is generally not a scalar value, so for demonstration,

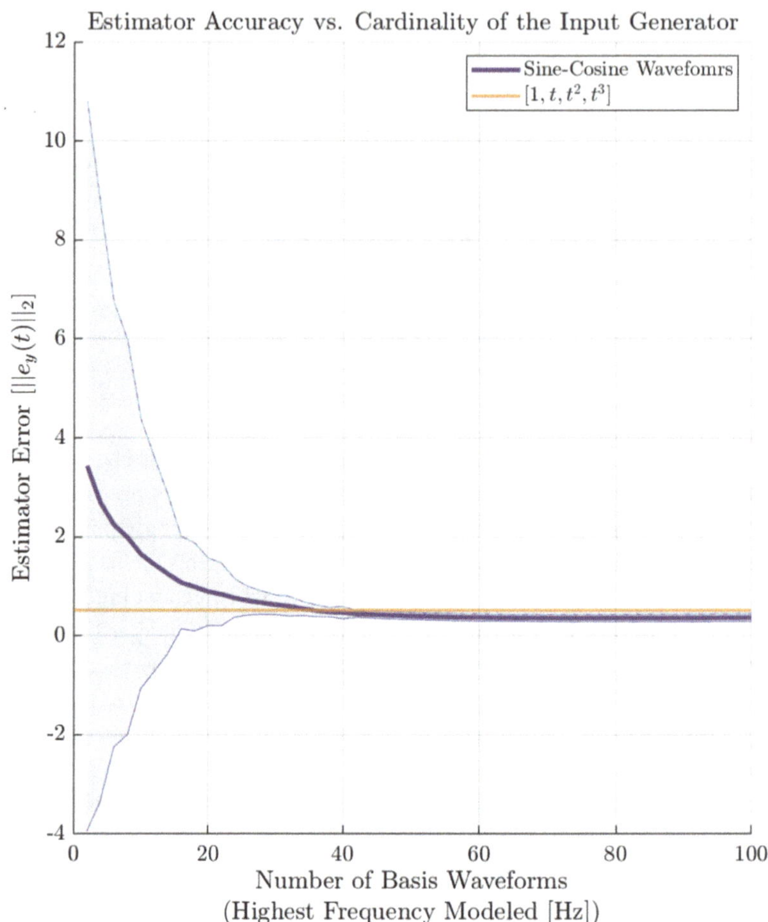

Fig. 6.3 UIO error as a function of basis generator cardinality. While the polynomial basis performs well and is computationally efficient, the sine-cosine basis is physically meaningful and eventually outperforms the polynomial estimate. (Reproduced from [11])

we may choose to look at the norm of the matrix $||L(t)||_2$ as an indicator of how important the adaptation of the system internal dynamics A_m is to the observed EEG data. A greater value of $||L(t)||_2$ should be interpreted as a point where the initial model synthesization is highly perturbed or where the data has nonlinear effects. Figure 6.6 demonstrates the time series of $||L(t)||_2$ for the same 60 s trial as was shown in Fig. 6.4. Here, this UIO shows particular strength over the simpler formulation in Eq. (6.2). By accounting for the external input and nonlinear behavior of the data, our estimate of the EEG wave is greatly improved. This is especially true when the EEG data has sudden spikes.

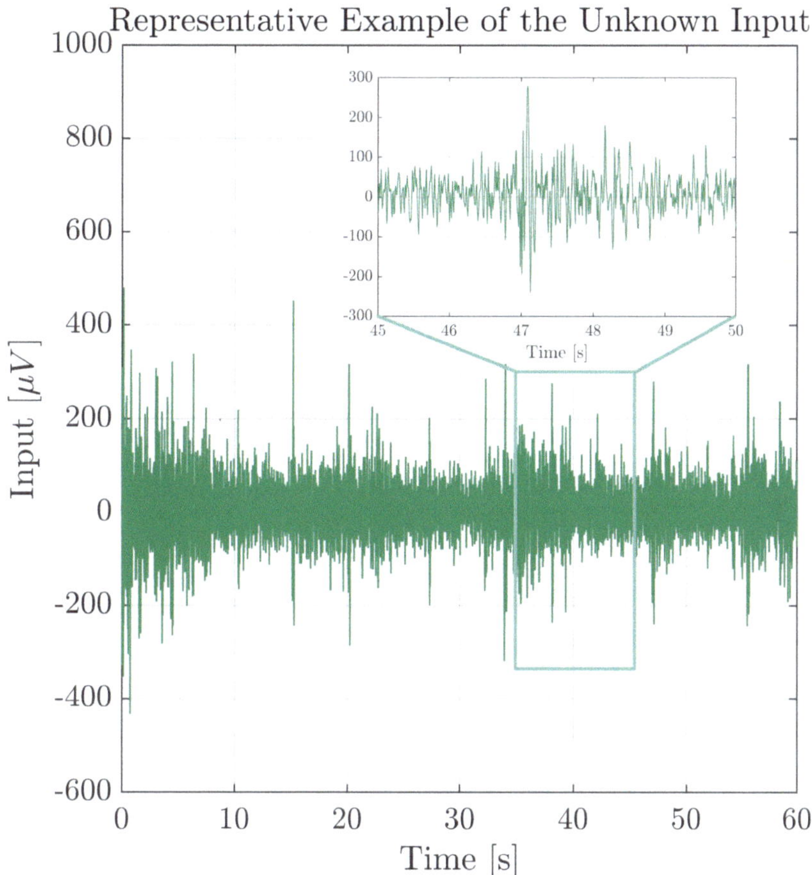

Fig. 6.4 A demonstrative example of the unknown input estimate for a single 60 s trial using sine and cosine waveforms. The input contains dynamics which cannot be generated by A_m. (Reproduced from [11])

Figure 6.5 is the key demonstration of this work. We show the time series of both our estimator and the standard Kalman filter in comparison to the true EEG recording. This is only a single channel in the 32 channel data, but it is representative of the whole. Although we saw in Fig. 6.2 that the *average* error of our UIO outperformed the Kalman filter, we see here that our UIO is especially effective in the important regions of the recording where there are significant spikes or other nonlinear effects. Notice that correspondingly, the adaptive gain becomes especially active at 42 s to compensate for this spike per Fig. 6.6. In this way, we may conceptualize $||L(t)||_2$ as another metric for the efficacy of the linear model. When the linear dynamics (A_m, C) are ineffective, the adaptive component of the estimator compensates for the observed dynamics. By treating the nonlinear behavior and compensating for uncertainty

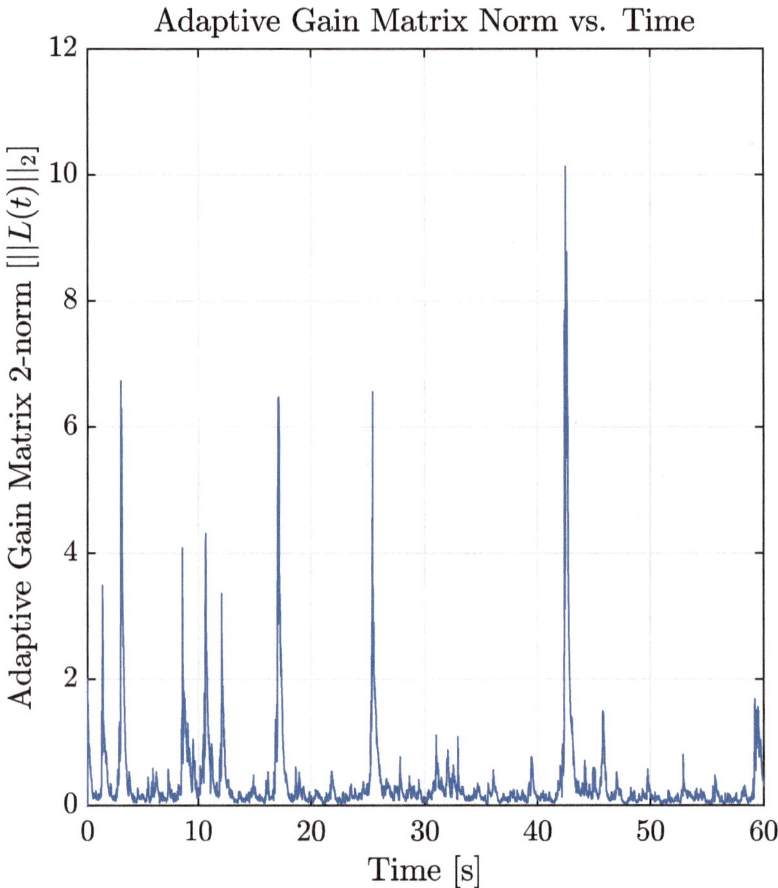

Fig. 6.5 A demonstrative example of the EEG predictions for a single 60 s trial compared to the true recording. While the estimator errors are similar over the entire time series, our UIO is especially effective in the spikes and other nonlinear regions of the data. (Reproduced from [11])

and the external input, our real-time estimate of the EEG data is greatly improved in the regions where the dynamics are especially active.

6.3.3   The Predictive Capability of the UIO

As a demonstration of the predictive capability of this unknown input observer, we present a binary classification task on two different datasets. In showing this result, we do not mean to imply this is the best or only way to draw analytical conclusions from the UIO. We do mean to imply that this demonstration shows that the UIO and all the work that led to its development

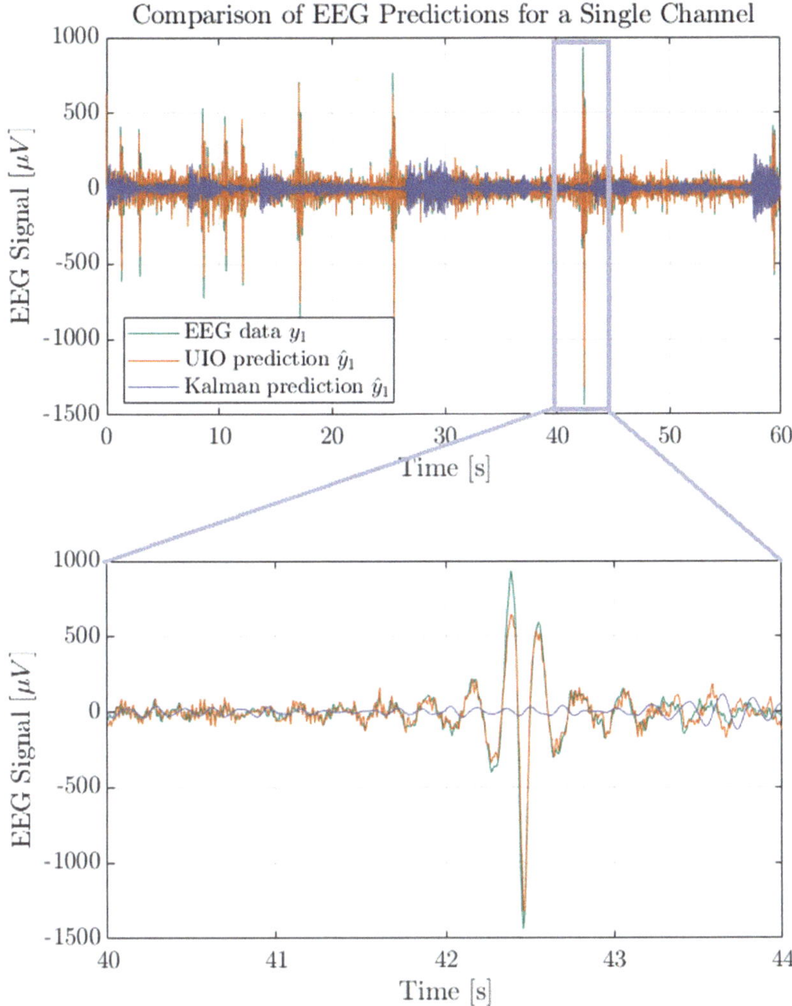

Fig. 6.6 A demonstrative example of the adaptive gain matrix for a single 60 s trial using sine and cosine waveforms. The estimator compensates for model error when the dynamics are highly perturbed or the data exhibits nonlinear effects. (Reproduced from [11])

yields a model that contains information relevant to analytical outcomes. In developing these models, we noticed that there are a number of shared eigenmodes within each individual subject. This suggests that there are stimuli independent and stimuli dependent modes as reported in Chap. 4. Accordingly, we hypothesized that eigenmodal decompositions from the same self- reported outcome should have more in common than decompositions from different self-reported outcomes.

Table 6.1 Comparison of other emotion identification results on the DEAP database

Reference	Valence accuracy (%)	Arousal accuracy (%)
[13]	57.6	62.0
[22]	71.05	71.05
[23]	78.2	74.92
[24]	81.40	73.4

Each of the 32 subjects in the DEAP database viewed 40 music videos which were selected for their ability to elicit strong emotions. Along with the biomarkers, the database includes self-reported ratings according to the valence-arousal scale [19]. This two-dimensional scale for emotional ratings is widely used as one of the repeatable emotional rating systems [20]. As with all self report measures, this scale is not perfect [21], but others have had success estimating the valence from this database per Table 6.1, where the accuracy is the binary classification accuracy of high or low valence and arousal. We note that the last three works cited above are deep learning or machine learning classification approaches. Because there are common eigenmodes within subjects, we hypothesized that an *averaged* view of the linear dynamics would reveal shared features among trials with the same self report. Therefore, the following procedure was empirically determined to classify valence from the EEG data:

1. Generate the pair (A_m, C) for 20% of the trials in the database for high and low valence for each subject (e.g. 8 models for high valence and 8 models for low valence in the DEAP database);
2. Average the generated models to determine an overall model for an individual subject (e.g. $(\bar{A}_H, \bar{C}_H)$ and $(\bar{A}_L, \bar{C}_L)$);
3. For each of the remaining unclassified trials, estimate each signal with the adaptive UIO for both the high and low averaged models; and
4. Predict high or low based on which estimator is more accurate over the trial.

The same procedure applies to high and low arousal. This process involves the simultaneous estimation of two different models. Figure 6.7 makes the process more clear, wherein the two averaged models estimate the same output data y. The estimator with a lower error over the entire trial $||e_y||_2$, more closely matches the unlabeled data, and so is more likely to have the same classification as that data.

While the estimator architectures are identical, we must find some deterministic way to place K_H and K_L. We could tune these gains so they improve our prediction accuracy, but that is somewhat arbitrary and would be an irregular estimation approach. Rather, we place the static gains K_H and K_L so they give the lowest error on the modeled data. We do this

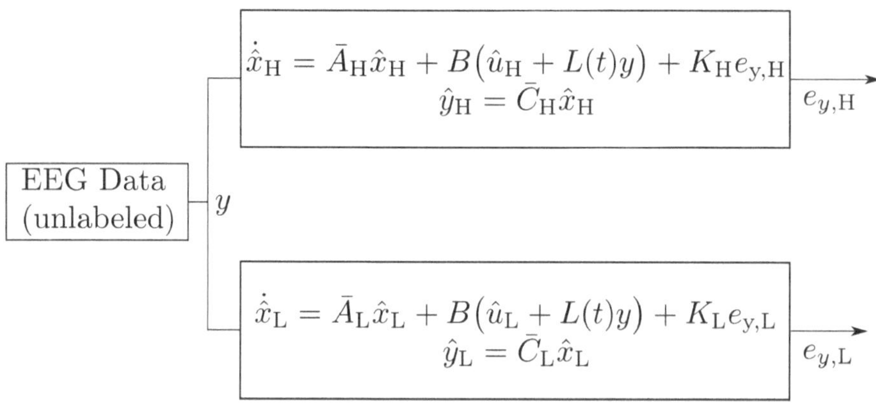

Fig. 6.7 Our approach to binary valence classification. By extracting averaged features for both high and low valence, we can label new data based on which averaged model is more accurate

Table 6.2 Our reported accuracy on unmodeled EEG data

Classification task	Estimator acc. (%)	PSD features acc. (%)
DEAP valence	77.8	68.1
DEAP arousal	75.2	63.8
Like/Dislike	79.4	67.3

with a coarse grid search over the 20% of the data that is modeled. For a continuous-time estimator such as this, the infinite-horizon cost function given by

$$J = \int_0^\infty [x^* Q x + u^* R u] dt, \tag{6.53}$$

where $Q = Q^* > 0$ and $R = R^* > 0$ yields the optimal control policy that minimizes the relevant Hamilton-Jacobi-Bellman equation

$$0 = \min_u [x^* Q x + u^* R u + \frac{\partial J^*}{\partial x}(A x + B u)]. \tag{6.54}$$

Our approach here is to vary Q and R by a scalar constant ranging from 1 to 100, increasing the constant by 10 after running the estimator over all the seen data. In the DEAP example, this yields 8 errors per estimator per cost function, for a total of 8×100 different combinations of Q and R. The gains which generate the lowest error over all the modeled data are selected for the predictive estimator in Fig. 6.7.

This modeling approach was applied to the DEAP database to classify high or low arousal and high or low valence. It was further validated on the Neuromarketing database to classify like or dislike from the unseen data. Table 6.2 shows the three fold classification accuracy for each of the classification tasks we considered in the middle column. For each fold, we randomly selected a different 20% of the data for initial modeling and averaging to make sure we were not biasing the estimation process.

As a means for relative comparison, the accuracy of a convolutional neural network (CNN) using windowed power spectral features is shown in the right column of Table 6.2. That is, the spectral density of each trial is calculated, and those features are provided to a CNN for each classification task. It can be seen that the approach in Fig. 6.7 outperforms this baseline spectral method, which suggests our approach is extracting some additional useful information from the raw data.

It is important to note that this estimation accuracy classification is an individualized model, not a global one. Each individual has two distinct estimators per Fig. 6.7. This is aligned with the notion that brain waves are inter-individualized, but it is worth exploring the possibility of a global estimator for the classification of unlabeled EEG data. For the neuromarketing dataset, a global estimator for all subjects was extracted by averaging the eigenmodes from all subjects in the database. The global model was seen to correctly classify the liked samples from the disliked samples with 52% accuracy on a three fold validation approach. Because this result was not encouraging, the global estimator approach was not extended to consider the DEAP database. However, since it has been demonstrated that the eigenmodes can be used to distinguish one subject from another in Sect. 4.3.1, this work recommends first that the subject be identified, and then the correct individually calibrated classification algorithm be implemented.

One may argue with the process of selecting static gains that give the "best" error. In Fig. 6.8, we show the number of correct classifications on the unmodeled data with randomly selected estimator gains. Boxed in red is our reported average accuracy. Since our classification accuracy by selecting gains that minimize the estimator error is lower than the average from a random gain, we conclude that this is a reasonable approach. Note, there are gains that yield more accurate classification outcomes, but we could only identify these gains by searching the unseen data for them randomly. There is now way to determine them from the modeled data, and it is our goal here to only adjust our classification procedure based on the modeled data.

Again, we wish to reinforce that this result merely validates our modeling approach. In the same way that the subject identification task in Chap. 4 validated our linear models, this classification task validates our adaptive UIO. There are vast other possibilities for classification here, including the use of the adaptive modes from $BL(t)C$ in conjunction with the original linear modes from A_m or the use of the unknown input, which we simply have not had time to explore yet in this modeling focused work. However, note that while the classification accuracy of the adaptive unknown input estimator in Table 6.2 is comparable to the existing approaches in Table 6.1, the analytical information generated with the adaptive unknown input observer is greater. Figure 6.9 more clearly shows this comparison. Curiously, the adaptive unknown input approach has greater accuracy on the arousal classification than on the valence classification, which is not observed in the other classifiers. Further, notice

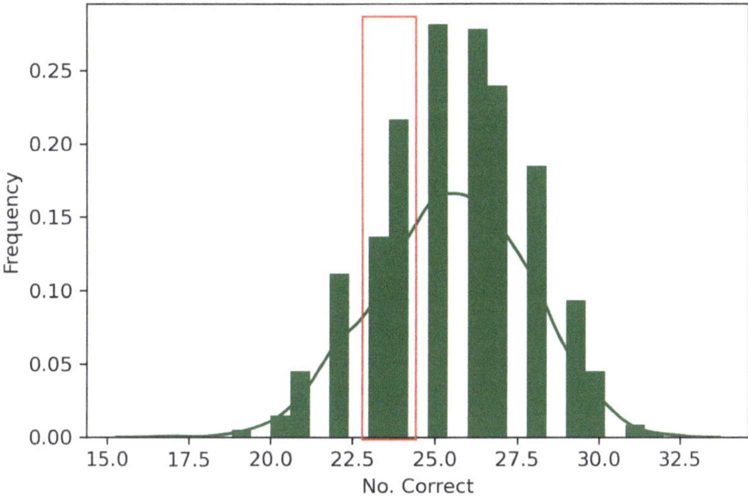

Fig. 6.8 Estimator accuracy when the static gains are chosen randomly. A perfectly accurate estimator would have 32 correct predictions. On average, our modeling approach, which places the static gains to minimize error, correctly labels 23 of the 32 validation EEG recordings

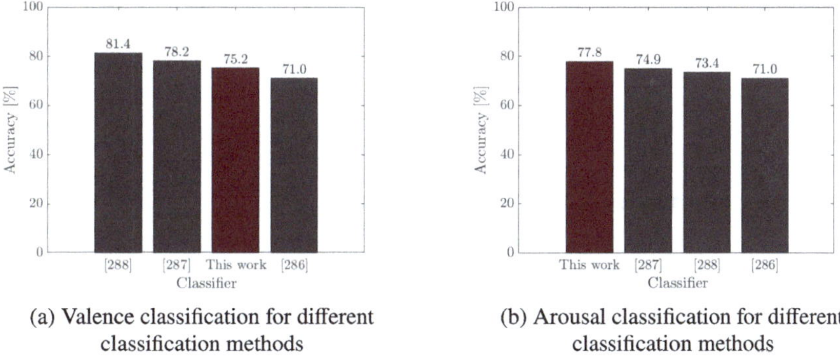

(a) Valence classification for different classification methods

(b) Arousal classification for different classification methods

Fig. 6.9 Comparison of adaptive unknown input DEAP classification with existing literature

the additional analytical information generated by the adaptive unknown input estimator. The unknown input estimate, adaptive gain matrix L, estimator error e_y, and brain wave eigenmodes are all available to help interpret the classification results.

6.3.4   Limitations of the Input Estimate

While we have an approach for the estimation of u, this estimator does not tell us anything about how the unknown input is distributed over the dynamics through the B matrix. In typical engineering systems, even if the input is unknown, we have some idea of how these perturbations act on the physical system. For the analysis of EEG data, it's unclear how the external input influences the individual channels. Because we have limited knowledge here, our approach is to estimate a single unknown input acting evenly over all channels. That is, we set B equal to a column of ones and generate one waveform $\hat{u}$. We could generate multiple unknown inputs, each with a different B matrix, but this view quickly becomes arbitrary. More importantly, the dynamics in Eq. (6.17a) can become unobservable when too many unknown inputs are estimated. While a single unknown input with $B = I$ may not seem better, we argue that this represents an averaged view of the unknown input which mirrors the smearing effect of EEG measures as a result of the cerebrospinal fluid, the skull, and the other disturbances in and EEG data. While this estimator is well suited to estimating multiple unknown inputs, as we show in [10], the waveforms need to be linearly independent and we do not have a good way to split the selected waveforms here into multiple different inputs. Analytically, we will show that this approach, while coarse, gives good results. Moreover, to our knowledge this is the first demonstration of an approach which yields an estimate of the input to a brain wave system. There are limits to how many unknowns we can tolerate in a given problem, and this assumption of a single input acting evenly over all channels is an example of this limit. There are adaptive estimators which can estimate the B matrix when the input is known, but this does not extend to the case when the input is unknown. When the B matrix and the input are unknown, it is mathematically impossible to separate between parametric error in the B matrix and error in the unknown input, so estimating both simultaneously is not possible.

6.3.4.1 An Optimization Routine to Improve B

Because we cannot update the B matrix in the presence of an unknown input, we explored optimization methods for improving the parametrization of B after the estimation process was completed for a given time series. We hypothesized that adjusting the B matrix so it minimized the error during the estimation process might yield physically significant insight into the unknown input and how it was acting on the brain. A convex function for minimization, in this case, was

$$\min \|y - \hat{y} - CB\hat{u}\|_2. \tag{6.55}$$

This convex function adjusts the values of B, while the other variables are constant. While it is not the only possible minimization function, it is a minimization function that is guaranteed to converge in polynomial time and it is guaranteed to have every local minimum be a global minimum. As a very simple example, consider the demonstration from Sect. 5.5, where we modify the true B matrix to

$$B = \begin{bmatrix} 1.2 \\ 1 \\ 1.6 \end{bmatrix}, \tag{6.56}$$

while our estimator is provided the column of ones as

$$B_m = \begin{bmatrix} 1 \\ 1 \\ 1 \end{bmatrix}. \tag{6.57}$$

We used CVX [25, 26] and fminsearch [27] routines to solve for an updated B_m using the objective cost function in Eq. (6.55). Figure 6.10 shows the output response y of the true plant compared to the optimized response and the unoptimized response. Note, while both CVX and fminsearch improve the estimate, it does not exactly recover the correct B. The identified B is

$$B_{m,f} = \begin{bmatrix} 1.18 \\ 1 \\ 1.37 \end{bmatrix}. \tag{6.58}$$

6.3.4.2 Optimizing the B Matrix for Brain Wave Estimation

We can apply this approach to the estimation of EEG brain waves. Rather than show a long 32×1 updated B matrix, we can image the change at each spatial location. Figure 6.11 shows an illustrative estimate of the unknown input along with the corresponding update to

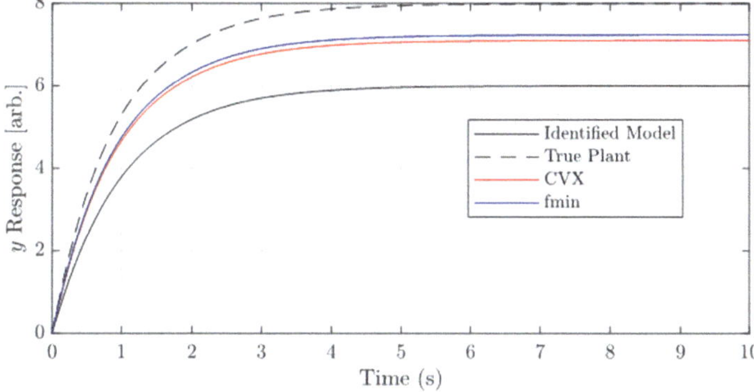

Fig. 6.10 A comparison of system responses for the true model, the original mode, and the model with an optimized B matrix. Two different optimization routines are compared. The optimization improves the estimate

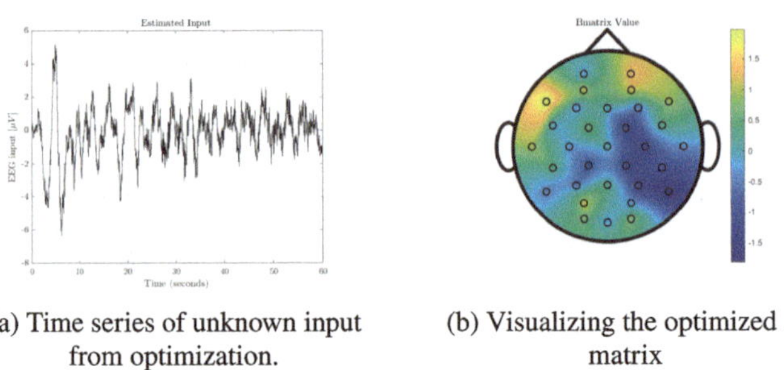

(a) Time series of unknown input
from optimization.

(b) Visualizing the optimized B
matrix

Fig. 6.11 Example of the results from optimizing the B matrix. Notice that some spatial regions have the effect of the unknown input increased, while others have it lessened

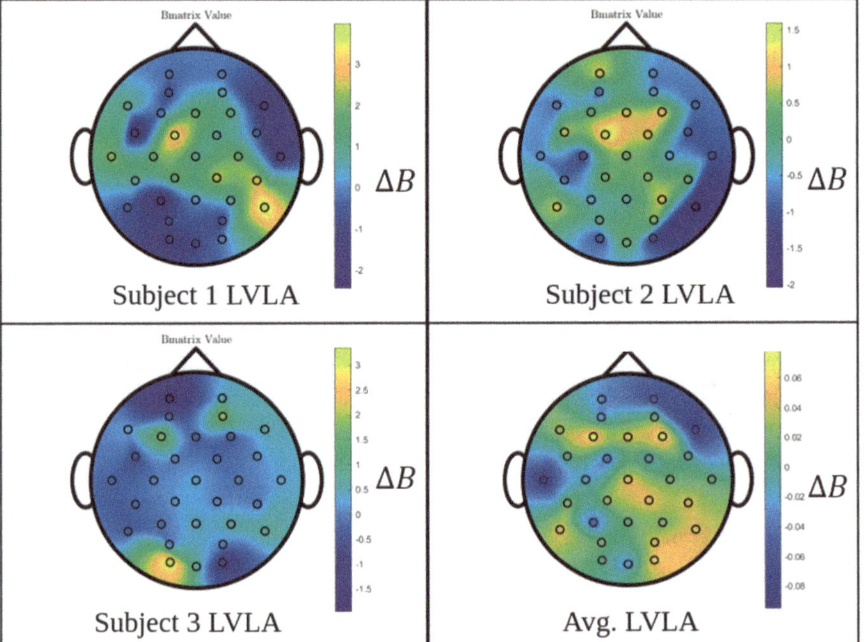

Fig. 6.12 Mapping the optimized B matrix for the low valence and low arousal quadrants. We were unable to estimate the self report from the spatial patterns in the optimized B

the B matrix. This spatial dependence of the B matrix was encouraging. We made significant attempts to identify common spatial patterns between the same self-reported valence-arousal quadrant in the DEAP database. Figure 6.12 shows the B matrix visualization for three subjects who all self-reported low emotional valence and low emotional arousal, along with the average B matrix for all subjects reporting low valence and low arousal. Notice, there are

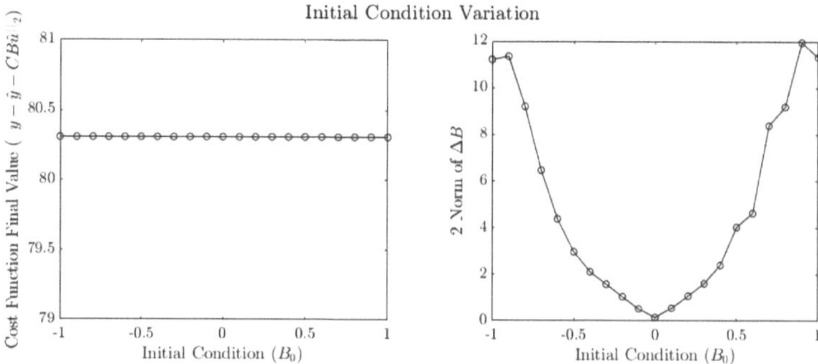

Fig. 6.13 The optimization of the B matrix has many solutions which yield the same minimum but have different parameterizations

few shared features between any of the four mappings. We were unable to classify emotional valence or arousal based on the B matrix optimization visually or with a convolution neural network.

Ultimately, while disappointing, this conclusion should not surprise us. Convex functions have the desirable characteristic of having all local minimums be global also. For this reason, the optimization of the B matrix in this way will be subject to the initial condition. Indeed, we determined that there are many B matrices that will yield a local minimum value of the cost function. Figure 6.13 shows exactly this. The plot on the left shows that the same minimum on the y-axis is achieved regardless of the initial condition on the x-axis, while the plot on the right shows that the two norms of the resultant B matrix varies greatly by initial condition. This problem *does not have a unique solution*, which makes classification difficult. However, it does further improve the fidelity of our models. Overall, however, the improvement is marginal compared to the computational complexity, so we generally do not include this step in our analysis of brain wave dynamics. OMA in combination with the adaptive state estimator, using a B matrix of ones, is sufficient to achieve excellent predictive capability. The introduction of the B matrix optimization routine takes away the benefit of real-time estimation, so we generally prefer not to include it in our modeling procedure.

6.4 Conclusions

In this work, we present the theoretical foundations for a real-time brain wave estimator that treats three key uncertainties in EEG data. While our estimate of the unknown input is hampered by uncertainty in the B matrix, we show that the performance of this estimator is excellent in comparison with standard linear methods. In addition, this estimator provides more information for the analysis of EEG recordings in the form of an estimate of the unknown input which compensates for the dynamics which cannot be generated by the model alone, and in the form of the adaptive gain matrix which compensates for nonlinear effects.

6.4.1 Recommendations

While the focus of this work was on the theoretical foundations of our estimator and modeling as a whole, we hypothesize that this approach offers much analytical insight. Correlating the unknown input with external stimuli is an especially appealing next step.

Overall, this novel adaptive approach to real-time estimation of EEG data demonstrates what to our knowledge is the first architecture to demonstrate a method of estimating an input in "brain wave space". It's tolerance to modeling errors, general uncertainty, and real-time implementation make it applicable to a wide variety of biomarker recordings.

References

1. C. Stam, Nonlinear dynamical analysis of EEG and MEG: review of an emerging field. Clin. Neurophysiol. **116**(10), 2266–2301 (2005), https://www.sciencedirect.com/science/article/pii/S1388245705002403
2. K.F.K. Wong, A. Galka, O. Yamashita, T. Ozaki, Modelling non-stationary variance in EEG time series by state space Garch model. Comput. Biol. Med. **36**(12), 1327–1335 (2006), https://www.sciencedirect.com/science/article/pii/S0010482505001083
3. B.L.P. Cheung, B.A. Riedner, G. Tononi, B.D. Van Veen, Estimation of cortical connectivity from EEG using state-space models. IEEE Trans. Biomed. Eng. **57**(9), 2122–2134 (2010)
4. H. Shen, M. Xu, A. Guez, A. Li, F. Ran, An accurate sleep stages classification method based on state space model. IEEE Access **7**, 125 268–125 279 (2019)
5. S. Miran, S. Akram, A. Sheikhattar, J.Z. Simon, T. Zhang, B. Babadi, Real-time tracking of selective auditory attention from M/EEG: a Bayesian filtering approach. Front. Neurosci. **12**, 262 (2018), https://www.frontiersin.org/article/10.3389/fnins.2018.00262
6. J. Roubal, P. Husek, J. Stecha, Linearization: students forget the operating point. IEEE Trans. Educ. **53**(3), 413–418 (2010)
7. C.-T. Chen, B. Shafai, *Linear System Theory and Design*, vol. 3 (Oxford University Press, New York, 1999)
8. P. Antsaklis, A. Michel, *A Linear Systems Primer* (Birkhäuser, Boston, 2007). https://books.google.com/books?id=7W4Rbqw_8vYC
9. M. Balas, S. Gajendar, L. Robertson, *Adaptive Tracking Control of Linear Systems with Unknown Delays and Persistent Disturbances (or Who You Callin' Retarded?)*. https://arc.aiaa.org/doi/abs/10.2514/6.2009-5855
10. T. Griffith, V.P. Gehlot, M.J. Balas, *Robust Adaptive Unknown Input Estimation with Uncertain System Realization*. https://arc.aiaa.org/doi/abs/10.2514/6.2022-0611
11. T. D. Griffith, V. P. Gehlot, M. J. Balas, and J. E. Hubbard, "An adaptive unknown input approach to brain wave eeg estimation," Biomedical Signal Processing and Control, vol. 79, p. 104083, 2023. [Online]. Available: https://www.sciencedirect.com/science/article/pii/S1746809422005444

12. V.P. Gehlot, M.J. Balas, Projection based robust output feedback direct adaptive systems, in *AIAA SciTech Forum* (2021)
13. S. Koelstra, C. Muhl, M. Soleymani, J. Lee, A. Yazdani, T. Ebrahimi, T. Pun, A. Nijholt, I. Patras, DEAP: a database for emotion analysis; using physiological signals. IEEE Trans. Affect. Comput. **3**(1), 18–31 (2012)
14. M. Yadava, P. Kumar, R. Saini, P.P. Roy, D. Prosad Dogra, Analysis of EEG signals and its application to neuromarketing. Multimed. Tools Appl. **76**(18), 19 087–19 111 (2017). https://doi.org/10.1007/s11042-017-4580-6
15. Q. Li, R. Li, K. Ji, W. Dai, Kalman filter and its application, in *2015 8th International Conference on Intelligent Networks and Intelligent Systems (ICINIS)* (IEEE, 2015), pp. 74–77
16. T.D. Griffith, J.E. Hubbard, System identification methods for dynamic models of brain activity. Biomed. Signal Process. Control. **68**, 102765 (2021), https://www.sciencedirect.com/science/article/pii/S1746809421003621
17. A. Gelb, *Applied Optimal Estimation* (MIT Press, 1974)
18. C. Johnson, Effective techniques for the identification and accommodation of disturbances, in *Proceedings 3rd Annual NASA/DOD Controls-Structures Interaction (CSI) Technical Conference* (1989), p. 163
19. T.M. Sutton, A.M. Herbert, D.Q. Clark, Valence, arousal, and dominance ratings for facial stimuli. Q. J. Exp. Psychol. **72**(8), 2046–2055 (2019), pMID: 30760113. https://doi.org/10.1177/1747021819829012
20. S. Mohammad, Obtaining reliable human ratings of valence, arousal, and dominance for 20,000 English words, in *Proceedings of the 56th Annual Meeting of the Association for Computational Linguistics (Volume 1: Long Papers)* (Association for Computational Linguistics, Melbourne, Australia, 2018), pp. 174–184. https://aclanthology.org/P18-1017
21. D. Watson, K. Stanton, L.A. Clark, Self-report indicators of negative valence constructs within the research domain criteria (RDoC): a critical review. J. Affect. Disord. **216**, pp. 58–69 (2017), rDoC Constructs: Integrative reviews and empirical perspectives. https://www.sciencedirect.com/science/article/pii/S0165032716307376
22. Y. Fang, H. Yang, X. Zhang, H. Liu, B. Tao, Multi-feature input deep forest for EEG-based emotion recognition. Front. Neurorobotics **14** (2021), https://www.frontiersin.org/article/10.3389/fnbot.2020.617531
23. M.R. Islam, M.M. Islam, M.M. Rahman, C. Mondal, S.K. Singha, M. Ahmad, A. Awal, M.S. Islam, M.A. Moni, EEG channel correlation based model for emotion recognition. Comput. Biol. Med. **136**, 104757 (2021), https://www.sciencedirect.com/science/article/pii/S0010482521005515
24. S. Tripathi, S.G. Acharya, R.D. Sharma, S. Mittal, S. Bhattacharya, Using deep and convolutional neural networks for accurate emotion classification on DEAP dataset, in *AAAI* (2017)
25. M. Grant, S. Boyd, Graph implementations for nonsmooth convex programs, in *Recent Advances in Learning and Control*, ed. by V. Blondel, S. Boyd, H. Kimura. Series Lecture Notes in Control and Information Sciences (Springer, 2008), pp. 95–110. http://stanford.edu/~boyd/graph_dcp.html
26. M. Grant, S. Boyd, CVX: Matlab software for disciplined convex programming, version 2.1 (2014), http://cvxr.com/cvx
27. J.C. Lagarias, J.A. Reeds, M.H. Wright, P.E. Wright, Convergence properties of the Nelder-Mead simplex method in low dimensions. SIAM J. Optim. **9**(1), 112–147 (1998)

Conclusions and Future Work

7

7.1 Summary

The joint analysis of spatio-temporal brain wave patterns is a developing topic in the engineering and neuroscience community. While the exact relationship between neuronal level activity and EEG data is unclear, useful clinical and diagnostic information has repeatedly been extracted from brain wave data. Because brain waves are not generated by any known first principles system, a wide variety of modeling techniques exist. Because the spatial dynamics are so important in the analysis of brain wave measures, this work proposes the development of a rigorous modeling procedure for brain waves in canonical state-space terms, which treats the brain wave process as a black box. Specifically, this includes the use of a modal representation of linear brain wave models, which are highly interpretable and physically significant. Of course, brain wave measures have significant nonlinear and nonstationary phenomena which must be accounted for in our modeling approach. From this background, Chap. 2 took a survey of the key modeling considerations as they related specifically to EEG data, including common corrupting artifacts and preprocessing methods.

To solve this problem, a variety of modal system identification methods were introduced in Chap. 3 to extract linear spatio-temporal patterns from EEG data. The relative performance of each technique was discussed and analyzed. Ultimately, the OMA and DMD algorithms were shown to extract highly consistent representations of brain wave dynamics. That is, they had the least variation in the extracted modes for a given subject's brain waves. Some initial analysis was presented, mainly the physical significance of the modes and a demonstration of how a superposition of the modes recreates the observed EEG data.

Further development of the modal approach to brain wave dynamics was presented in Chap. 4, where technical details of the OMA and DMD algorithm were presented. Analysis of the EEG modes was shown, including the existence of subject-independent common modes, which are spatial patterns at distinct frequencies appearing in every subject of the

  127
T. D. Griffith et al., *A Modal Approach to the Space-Time Dynamics of Cognitive Biomarkers*, Synthesis Lectures on Biomedical Engineering, https://doi.org/10.1007/978-3-031-23529-0_7

DEAP database. While these EEG modes only model the linear behavior of brain wave dynamics, it was shown that they are sufficient to distinguish one subject from the next with perfect accuracy using a machine learning classification approach.

Following this analysis, additional work was undertaken to model the nonlinear phenomena and parametric uncertainty associated with brain waves. To solve this problem, a general architecture for the treatment of nonlinear plant dynamics while estimating the unknown external system input was presented in Chap. 5. This novel estimator was designed specifically to update the brain wave model in real time to account for nonlinear or nonstationary effects through the implementation of an adaptive gain law. Additional proofs for a robust adaptive unknown input estimator and an adaptive unknown input estimator are presented in the appendices.

In Chap. 6, the adaptive unknown input estimator was applied to EEG recordings. The efficacy of the approach was demonstrated and the resultant models were analyzed. This novel estimator was shown to outperform other common estimators, especially when nonlinear effects were obvious in the data. The adaptive unknown input estimator was used to categorize subjective valence and arousal ratings from subjects in the DEAP database with comparable accuracy to existing deep learning approaches.

7.2   Key Contributions of This Work

7.2.1   Modal Identification of Linear Brain Wave Dynamics

Brain wave recordings via EEG are known to have stochastic, nonlinear, and nonstationary dynamics. As a result, there are significant complexities associated with the synthesization of accurate dynamical models from EEG recordings. This work presented a structured, canonical modeling pipeline to extract such models from EEG data. Accordingly, the first contribution of this work is the analysis and validation of modern system identification techniques to extract modal decompositions from EEG data. These dynamic modal decompositions represented observed data in a discrete set of spatiotemporal patterns that may be superposed to recreate the original data. These modes are amenable to the imaging and analysis of dynamic brain wave activity. Because the input to the brain wave plant is unknown, a significant portion of this contribution involved methods for identifying the number of modes necessary to describe a given EEG recording, and how those modes may be truncated. Ultimately, this work demonstrated that over the right time interval, brain wave dynamics may be described by modes that are separable in space and time.

7.2.2   Analysis of Spatio-Temporal Brain Wave Modes

While not all of the brain wave dynamics were captured with the linear state-space models, the models still had useful analytical information. This work demonstrates a number of analytical findings in the modal representation of brain waves, including that the brain wave modes are highly inter-individualized. In particular, a set of common modes were observed in subjects at distinct frequencies. That is, in the data sets analyzed for this work, there are task independent spatial patterns that vary from subject to subject. This reinforces the notion that future brain wave models must be inter-individualized. Additional analysis demonstrated methods for reducing the number of EEG electrodes needed for a given classification task and demonstrated the physical significance and interpretation of the brain wave modes.

7.2.3   Theoretical Considerations for the Estimation of Unknown Inputs

The work further contributes a significant body of knowledge to the theoretical underpinnings of unknown input estimators. In particular, an adaptive observer approach was taken to simultaneously recover the physical structure of the internal plant dynamics and the unknown input waveform. Sensitivity to general process and sensor noise was further investigated, resulting in a robust version of the unknown input estimator. Finally, the stability proof was further extended to treat known nonlinear plant dynamics robustly. While the input can be unknown, the engineer must select a basis of waveforms that is believed to comprise the unknown input. Much analysis was dedicated to evaluating different and less than optimal bases for the unknown input estimator.

7.2.4   Online Estimation of Nonlinear Brain Wave Dynamics

The developed robust adaptive unknown input estimator was applied for the modeling of nonlinear brain wave dynamics. It was demonstrated that by compensating for nonlinear effects, the adaptive unknown input architecture could estimate brain wave recordings in real time. By updating the modes while recording the data, an improved model fit is obtained while the physical interpretability of the model is retained. A surprising outcome here was that the estimator converges even if the model is initialized with a perturbed set of modes. This estimator was used in a binary classification task on the DEAP database. This binary classification was further validated on a second dataset. The resultant accuracies were comparable to those of the existing deep learning approaches.

7.3   Recommendations for Future Work

Many additional questions were generated during the preparation of this work. Following are some of the main questions that are left unexplored.

7.3.1   Considering Multiple Data Types

Much of the discussion on modes and neural dynamics has thus far been constrained to EEG measures. However, OMA and DMD can extract modes from any output vector y. Any set of biomarkers could be included in the output y for modal decomposition. Moreover, multiple biomarkers could be analyzed. With only EEG data, the generated eigenvectors capture the spatial relationship between each channel for a given frequency λ_i. If another biomarker was included, such as EKG measures, the extracted modes would inform the relationship between the two measures, because the mode shape will include all available channels in y. This is common practice, for example, in flight dynamics, where the joint interaction of roll, pitch, and yaw are controlled with mode shaping [1].

7.3.2   Improved Diagnostics and Analysis

This work was primarily focused on the task of developing high-fidelity real-time models of brain waves. A huge amount of potential exists for clinical and modeling diagnostics. For example, a serious exploration of the analytical information in the unknown input was not undertaken. Certainly, by the improvement in the model, there is exogenous information to the brain wave plant which is captured in the estimate of the unknown input. One of the key struggles of this work was interpretation. Additional insight from the neuroscience community is needed to truly understand modeling outcomes. As this is an engineering document, there is little sense in drawing clinical conclusions from our models. Accordingly, the analysis here was restricted to modeling and estimation. However, this technique offers a spatio-temporal online model of neural activity, which should be relevant to the outcomes related to neuroscience.

7.3.2.1 Accessibility of Modeling Procedure

The modeling pipeline of generating a good linear brain wave model and then using an adaptive estimator to correct the model in real time is well described in this work. However, recreation of the results would require an experienced graduate engineer. All of the modeling hyperparameters (system order, estimator poles, number of Hankel matrix rows) are tied to some rigorous engineering metric that determines their value.

In order to achieve widespread implementation in clinical trials, this modeling procedure would need to be increasingly automated. For example, the number of rows in the Hankel matrix should exceed the ratio of system order to the number of samples in the time series data, so automating the selection of the Hankel rows is possible. Improved packaging of the pipeline is necessary for widespread use.

7.3.3   Spatial Filtering of Biomarkers

Biomarker analysis generally suffers from the existence of many different filtering and feature extraction methods. Data collection and processing methods vary from lab to lab, which may impact modeling outcomes. This work has demonstrated a high-fidelity method for the real-time estimation of general biomarker data, but it will model whatever data is provided. If there are corrupting noise sources or filtering artifacts, these will appear in the identified eigenmodes. More work is needed to analyze the consequences of filtering on the provided biomarker data.

In order to preserve as much useful analytic information as possible, the approach in this work was to minimally process the EEG data. It may be possible to identify and correct common corrupting sources, such as ocular artifacts, by analyzing the identified eigenmodes. In this work, such artifacts were generally accepted as part of the system, because they would be present in the real-time measurement. Offline analysis may improve the feature extraction and modeling outcomes.

7.3.4   Probabilistic Considerations

From the analysis of common brain wave modes, it is clear that individualized models of brain wave dynamics are necessary. This work did not explore how much different external stimuli perturb the uncommon modes. Analysis in this area may indicate which regions of the brain are more or less influenced by different tasks, such as EEG.

More importantly, we note that this work demonstrated a highly efficacious model of brain waves. The dynamics of EEG recordings are captured by this process. However, the classification accuracies on cognitive modeling outcomes, such as valence and arousal for the DEAP database, left something to be desired. Since the dynamics are well modeled and the classification is only moderate, it may be concluded that human cognition is a probabilistic process. The same brain wave dynamics may not give rise to exactly the same self- reported cognitive outcome. More importantly, humans may be unreliable sources of self-investigation. This suggests that a probabilistic approach to brain wave dynamics may be useful for improved modeling outcomes such as attention or situational awareness.

Reference

1. B.L. Stevens, F.L. Lewis, E.N. Johnson, *Aircraft Control and Simulation: Dynamics, Controls Design, and Autonomous Systems* (Wiley, 2015)